An American Aristocracy

An American Aristocracy

Southern Planters in Antebellum Philadelphia

Daniel Kilbride

The University of South Carolina Press

Published by the University of South Carolina Press
Columbia, South Carolina 29208

www.sc.edu/uscpress

Manufactured in the United States of America

15 14 13 12 11 10 09 08 07 06   10 9 8 7 6 5 4 3 2 1

Library of Congress Cataloging-in-Publication Data

Kilbride, Daniel, 1968–
An American aristocracy : southern planters in antebellum Philadelphia / Daniel Kilbride.
p. cm.
Includes bibliographical references and index.
ISBN-13: 978-1-57003-656-9 (cloth : alk. paper)
ISBN-10: 1-57003-656-X (cloth : alk. paper)
1. Philadelphia (Pa.)—Social life and customs—19th century. 2. Philadelphia (Pa.)—Social conditions—19th century. 3. Interpersonal relations—Pennsylvania—Philadelphia—History—19th century. 4. Aristocracy (Political science)—Pennsylvania—Philadelphia—History—19th century. 5. Aristocracy (Political science)—Southern States—History—19th century. 6. Plantation owners—Southern States—History—19th century. 7. Leisure class—History—19th century. 8. Social classes—Pennsylvania—Philadelphia—History—19th century. 9. Community life—Pennsylvania—Philadelphia—History—19th century. 10. Sectionalism (United States)—History—19th century. I. Title.
F158.44.K55 2006
974.8'1103—dc22   2006019671

This book was printed on Glatfelter Natures, a recycled paper with 50 percent postconsumer waste content.

Contents

Illustrations

Acknowledgments

I am delighted to take this opportunity to recognize the many debts I have accumulated since I began this project around 1993.

This book began at the History Department of the University of Florida. The combination of a first-rate faculty, a friendly and efficient staff (particularly Betty Corwine), and a congenial cohort of graduate students made the department a special place in the 1990s. I thank the department for its generous financial support. I must single out Ronald P. Formisano and the late Darrett Rutman for embodying a peculiar combination of intellectual rigor and congeniality. This book—and more importantly my work as a historian, teacher, and mentor—has benefited incalculably from their example. Thomas Gallant, Jeffrey Adler, and John Seeyle of the English Department also read the manuscript and proffered candid advice about how it might be improved.

I owe my greatest obligation to Bertram Wyatt-Brown. As anybody who knows him can attest, Bert is much more than an academic advisor. More than anyone else, he fashioned the atmosphere of Florida's History Department in the 1990s, creating a place where congeniality, not competition, reigned; where academic rigor, not pretension, was the standard. Of course there were the intangibles: the parties, volleyball, travel support, fashion and culinary advice, and the rest. Thank you, thank you, thank you. Warm thanks also go to the group of graduate students with whom I studied (among other things), particularly the cohort who entered the department with me in 1990. I have cleverly disguised their names to protect the innocent: Granddad, Dutch, Gomez, Ignatius, and Brains. To this list I also add those who need no secret identity: Lisa Tendrich Frank, Andrew Frank, Tim Huebner, Kermit Hall, Jay Malone, Anne Jones, Stephanie Cole, Dan Stowell, and Christopher Morris.

Several people have read all of or parts of this manuscript, or what eventually became this manuscript, and I have benefited from their feedback. In no particular order, I recognize Joan Cashin, Peter Kolchin, Jan Lewis, Richard Bushman, Michael O'Brien, Charlene B. Lewis, and Christine Jacobson Carter for their kind and thoughtful advice. A number of institutions provided me with funding, office space, and other kinds of support. A year spent at the Philadelphia Center for Early American Studies (now the McNeil Center) provided me with an office, computer resources, and, most importantly, companionship during a crucial part of this project. Special thanks go to Richard Dunn, the center's director at that time, for his hospitality. I would also like to single out Simon Newman, George Boudreau, Cynthia Van Zandt, Billy Smith, and particularly Kirsten Wood for their fellowship. Mellon Fellowships from the Library Company of Philadelphia and Virginia Historical Society provided me with the means to conduct research in their unparalleled collections. I thank their staffs and also those at the Southern Historical Collection at the University of North Carolina, the Historical Society of Pennsylvania, Duke University's Perkins Library, the American Philosophical Society, the University of Pennsylvania, the Mississippi Department of Archives and History, and the Tompkins-McCaw Library at the Medical College of Virginia. I also thank my in-laws, George and Kathy Markwalter, for providing sustenance, shelter, and hospitality at their home in Greensboro, my jumping-off point for research in southern archives. They made my life a lot easier.

In very different form, several chapters or parts of chapters have been published as articles. Parts of chapter 1 were utilized in "Cultivation, Conservatism, and the Early National Gentry: The Manigault Family and Their Circle," *Journal of the Early Republic* 19 (Summer 1999): 221–56. An early version of chapter 6 appeared as "The Cosmopolitan South: Privileged Southerners, Philadelphia, and the Fashionable Tour in the Antebellum Era," *Journal of Urban History* 26 (July 2000): 563–90. The same is true of chapter 4, previously known as "Southern Medical Students in Philadelphia, 1800–1861: Science and Sociability in the 'Republic of Medicine,'" *Journal of Southern History* 65 (November 1999): 697–732. Parts of chapter 2 and the epilogue were published in "Class, Region, and Memory in a South Carolina—Philadelphia Marriage," *Journal of Family History* 28 (October 2003): 540–60. I thank the editors of these journals (Michael Morrison, David Goldfield, John Boles, and Joan Cashin, who edited a special issue, respectively) and the anonymous commentators who read the manuscripts for their assistance.

This book is dedicated to my parents and my grandparents. It is their accomplishment at least as much as it is mine. My wife, Heather, and our children Lauren, Andrew, and Shannon showed up later but have also indelibly left their mark on it and me.

An American Aristocracy

Introduction

IN HIS *Creating an Old South: Middle Florida's Plantation Frontier before the Civil War,* Edward Baptist observed, "Too often, histories of the nineteenth century South have looked like a cascade of inevitability. . . . The Civil War comes, right on time. Such histories become monologues trooping to a foreordained end." It is a difficult trap to escape; the knowledge that the Civil War did, in fact, occur inevitably colors our appreciation of the events that preceded it. The temptation to see the history of the prewar decades as a series of mounting antagonisms is powerful. And should it be resisted altogether? Arguably the most important event in the history of the United States, the Civil War demands explanation. It seems only fair that events unrelated to its origins surrender pride of place to those that were central.[1]

Nevertheless, this book takes a different tack. One important school of Civil War causation—one that, to judge from my students' attitudes, enjoys wide authority over the popular mind—argues that the North and South clashed because they were so different in so many ways that they could not exist peacefully within the same nation.[2] How to test such a proposition? One is drawn to extreme statements of regional difference: southern cavaliers versus northern Yankees; Norman southerners confronting Saxon northerners; a feudal South struggling to resist the dynamic North with its satanic mills, and so on. In fact, there is much evidence to sustain such an interpretation: northerners and southerners were given to caricaturing each other in newspapers and magazines, on the stage, and in virtually every other antebellum media.[3] To examine the extent to which the political struggle over the extension of slavery into the territories bled into the wider culture, *An American Aristocracy* examines concrete relationships between northerners and southerners in a specific time and place: early nineteenth-century Philadelphia. It asks a series of simple questions: how well did they get along together; how did the temper of their relations change (or remain stable) over time; and why did they do so? Because travel over such long distances was difficult and expensive, the southerners in question were, for the most part, major slaveholders. The women and men with whom they circulated in Philadelphia were correspondingly wealthy. Their interactions took place in suitably refined places: exclusive schools, ballrooms and salons, learned societies, and the like.

Because these interactions took place among people who were almost always privileged, this book examines social class as much as it assesses sectional identity. The two forms of self-identification, it turns out, had an interesting relationship. Planters and Philadelphians saw each other far less as southerners and northerners than as fellow aristocrats. This term may seem inappropriate. People of extraordinary wealth or privilege had no special legal standing in the United States, but the term accurately represents the ideal to which many socially prominent, conservative Americans aspired in the early decades of the republic. As one Philadelphia gentleman asserted, "in all civilised societies, an aristocracy must and will exist, either founded on letters, family, or fortune: it is either a *de jure,* as existing in Great Britain, or a *de facto* aristocracy, as existing in this country; and the power thus enjoyed by blood, by riches, and by learning, is as extensively exercised, and produces as great an effect over the minds of the lower orders of people."[4]

This ideal had never been realized in the colonies, and after the Revolution it was openly reviled as anachronistic and un-American.[5] Privileged Philadelphians and their southern friends and relations did not aspire to become a formal aristocracy—an impossible goal—but they did believe that persons who enjoyed wealth, sophistication, education, exposure to the wider world, and other privileges ought to receive deference from those who did not. Northerners and southerners who saw themselves in this way recognized important differences between their regions, of course. But these differences did not, at least until the late antebellum period, seem particularly important or threatening. For a long time class united more than section divided. In addition socially prominent northerners and southerners both felt under attack by the postrevolutionary demonization of elitism and, later, by the rise of the middle class. Not only were middling people powerful and numerous, but they challenged their superiors in an ingenious manner. Instead of repudiating the gentry's code of behavior, they co-opted it, rewriting the aristocratic ethic of gentility to suit their own needs.[6] Both northern and southern gentlepeople were left reeling by this multifronted assault.

Our understanding of the class dynamics of the early republic needs to accommodate the presence of this leisure class. Its significance has been obscured by social historians' desire to "uncover the once-hidden history of 'common people' and to de-emphasize those who for so long had dominated historical narratives."[7] Nor will it do to define membership in the privileged caste by wealth, since upper classes in American cities were divided along a myriad of confusing, overlapping divisions based on religion, family, etiquette, and many other qualities.[8] A devotion to an ever-more anachronistic aristocratic culture bound gentlepeople in the South and Philadelphia into an American leisure class during the first half of the nineteenth century.

Privileged, conservative people could be found in every region of the nation, but Philadelphia was renowned for the reactionary sensibility and aura of entitlement that distinguished its fashionable society. Thomas Hamilton, a Scot who traveled through the United States in 1830, noted, "There is no American city in which the system of *exclusion* is so rigidly observed as in Philadelphia. There is a sort of holy alliance between its members to forbid all unauthorized approach. Claims are canvassed, and pretensions weighed; manners, fortune, tastes, habits, and descent, undergo a rigid examination." The city was also close to the South, both geographically and temperamentally. Planters traveling to New York, Boston, or other points north usually traveled through Philadelphia first, guaranteeing that the city's boardinghouses and hotels were bursting with southerners. The city also had a southern orientation. Sympathetic observers and critics alike detected a strong affinity between its fashionable society and the southern gentry. When sectional tensions arose, these tendencies only intensified, lending to the city a decided southern air. The journalist Charles Godfrey Leland, who had grown up amid this society, bitterly remarked that "everything southern was exalted and worshipped" among the Philadelphians of his family's acquaintance.[9]

Privileged Philadelphians and their southern peers were strongly inclined to like and admire each other. This affinity rendered them strongly resistant to centrifugal forces—such as sectional partiality—that threatened their community. In 1835 the Philadelphia gentleman Joseph Hopkinson (author of "Hail Columbia") offered his praise to the editors of the *Southern Literary Messenger,* noting that such publications "tend to remove prejudices excited by vulgar anecdotes and the practices of vulgar men." The *Messenger* would help to "show that there is no important difference of character, education or habits, between gentlemen of the same grade in the South and the North. Each have the same peculiarities in their modes of life," Hopkinson admitted, "but none of them affect the substantial ingredients of their personal and national character." An examination of relations between southerners and northerners in Philadelphia affords the opportunity to see how sectional tension affected a community that was not only free of its taint, but actively resisted it.[10]

1

The Carolina Row

In Susan Petitgu King's 1855 novel *Lily,* Alicia Barclay takes the heroine, a new arrival at the Philadelphia boarding school they both attend, for a stroll through the city. Passing by a row of houses, the older girl informs Lily that in days gone by the homes "were called Carolina Row. . . . such numbers of our country people occupied and owned them." King—herself educated at one of the city's finishing schools in the 1820s—was the daughter of South Carolina Unionist James Louis Petigru. Alarmed by the sectional antagonism of the 1850s, she may have been anxious to remind her readers of a time when northerners and southerners, driven to construct a republic rather than to tear one down, enjoyed each other's company. If such was her intention (and she certainly shared her father's pro-Union sympathies), she did well to draw attention to a group of nondescript red brick row houses.[1]

Alicia and Lily may have been fictional, but the Carolina Row was real. It was a second home for a colony of South Carolinians who purchased houses adjacent to one another on Spruce Street between Ninth and Tenth, a fashionable district of downtown Philadelphia opposite Pennsylvania Hospital. It became famous as a social center where conservative Americans and Europeans could gather for conversation and entertainment in an ideologically congenial atmosphere. The Carolina Row came about when Margaret Manigault, a South Carolina widow, decided to establish a headquarters of sorts for her far-flung family. The Manigault family had long been among the wealthiest and most-prominent clans in South Carolina, but by the early nineteenth century its various branches were scattered across New York, Pennsylvania, and continental Europe. Manigault was more than a leader for her family. She was also a committed Federalist partisan and an elitist—a reactionary, even—who wished to take a stand against what to her seemed to be the alarming progress of democratic sentiment in the young nation. To succeed she needed her family around her. She also needed a central location from which her efforts could resonate across the nation. Hence a provincial center such as Charleston or Boston would not do. It had to be Philadelphia.[2]

Though she was de facto leader of her family after the death of her husband, Gabriel, in 1809, Margaret Manigault did not make the decision to relocate her family from South Carolina to Philadelphia alone. Her mother, Alice DeLancey Izard, had been a champion of the city for a number of years. Alice's motivations also stemmed from family concerns. She missed the company of her eldest daughter and her grandchildren. Since many of the latter—particularly the girls—were approaching adolescence, she believed they needed a grandmother's supervision. But she was also an active partisan, a conservative at a time when Federalists began to see the future slipping away from them. About 1800, when Thomas Jefferson had ascended to the presidency, women and men such as the Manigaults began to be edged out of the young nation's cultural mainstream. Though men of wealth continued to hold sway in the republic's councils (and did, moreover, for a long time to come), they were not assuaged. They were not satisfied with influence. They wanted power: direction not merely over the nation's political institutions but over its cultural tone and social relations. Thus, when in 1811 Alice complained of the "present riotous, boisterous manners," she was not making a petty complaint, but a pointed critique about the declining influence of conservatives in the young United States.[3]

For reactionaries such as Alice Izard, the evidence was all around, too obvious to question: the rise of a rabidly anti-aristocratic press, lingering sympathy for the French Revolution and—worse—adulation for Napoleon, the eclipse of the Federalist Party, insolence from servants. It all added up to a shocking decline in deference from the lower orders toward their social superiors, which seemed to have accelerated since Jefferson's election. It is easy to dismiss their fears as hyperbole. Compared to England and to continental Europe, deference had been weak in all the mainland colonies even before independence.[4] Yet to wellborn women and men, the colonial era looked almost like an Elysium compared with the early national era. Emboldened by revolutionary rhetoric, which reinforced an exaltation of human rights with a blistering critique of elitism, ordinary people openly disparaged privileged folk as dissipated, self-absorbed ciphers. Reactionaries' fears were neither entirely exaggerated nor wholly self-interested.[5] Certainly, they overstated the extent of their influence in former times. But they sincerely believed that the republic was the worse for their marginalization. Without their steady influence, it would careen inevitably toward mob rule, with all the implications of the Place de la Concorde. Cultural life would become vulgar and dissipated. Society, bereft of moorings, would suffer the wholesale elimination of all standards of good taste. As far as Alice Izard and Margaret Manigault were concerned, at stake in the Carolina Row was nothing less than the future of American civilization.

As privileged women, mother and daughter assumed that social leadership, including political influence, was their birthright. As early nationals, they saw the struggle between privilege and democratization as an ongoing struggle, not as history. And as members of a cosmopolitan family they waged cultural warfare as Americans, not as southerners. All three of those conditions were unique to the Manigaults' time and place. Women in subsequent decades had to (and did) struggle against a powerful ideology of domesticity in order to participate in the public realm. In postrevolutionary decades, however, women used their control of social affairs in Washington and elsewhere to influence the process of nation building, to control access to power, and to define the standards of respectability by which political actors were judged. Thus, though they enjoyed little influence on public policy, early republican women did mold other aspects of political life, especially caucusing, consensus building, public opinion, and patronage.[6]

The Carolina Row's women sought to steer the young republic's culture in precisely this way. They set out to reinvigorate the Federalist Party, reeling from successive defeats and humiliations in the post-1800 period. The threat faced by Jefferson's hordes was not merely political. Rather, their rise assaulted core national values and threatened to undermine Federalist claims that they alone could defend the nation against threats foreign and domestic.[7] Smarting from Charles C. Pinckney's drubbing at the hands of James Madison in 1808, Alice Izard pined for "some unexpected event [that] should suddenly open the eyes of the blind." But she was not content to wait for a miracle. Izard and her daughter hit upon a more practical expedient: they made their homes into the center of a Federalist social circle. They conceived of it not as a retreat from an increasingly hostile political culture, but as a bulwark against it. Believing that American aristocrats suffered from a failure of nerve, the Carolinians fashioned a salon where reactionary sentiments could be uttered openly, where quasi-aristocratic social forms could be practiced unapologetically. These women did not see culture and politics as discrete categories; aristocratic styles of dining, dance, and conversation reinforced elitist social principles and political ideals. The women of the Carolina Row thereby blurred the line between private and public, though that line was not as clear in the early national era as it became later on. What distinguishes them from the social elites of the Jacksonian era is that they refused merely to pine for past glories or hope for a deus ex machina to revive their fortunes—they acted.[8]

". . . a comfortable house in town"

Philadelphia possessed a number of virtues in the Manigaults' eyes, but in one respect their decision to settle there was very simple: many Carolina planter families enjoyed long-standing and warm associations with the city. A year after the British captured Charles Town in May 1780, they sent nearly two hundred patriot leaders and their families to Philadelphia in a prisoner exchange. The northerners greeted their fellow patriots warmly. Josiah Smith, one of the distressed Carolinians, testified that several "persons gave the use of Country houses to the exiles until they cou'd accommodate themselves more comfortably in the City &c and a number of them were kindly entertained by Gentlemen &c in their City dwellings." Philadelphians augmented a congressional loan with $15,000 in credit and more than $3,300 in donations. It is likely that some tensions marred relations between the South Carolina exiles and their northern hosts. It could hardly be otherwise, however, as war-weary Philadelphians struggled to accommodate several hundred distressed refugees. As the Carolinians prepared for their return home in the winter of 1782–83, however, memories of good fellowship soothed raw feelings. Josiah Smith closed his refugee diary by paying tribute to an "act of kindness" performed by the city's Second Presbyterian congregation. It had honored his minister father by interring his body in the middle aisle of its church, next to Gilbert Tennent, the renowned revivalist.[9]

Between 1790 and 1800 Philadelphia served as the capital of the young republic. It became notorious as the seat of the "Republican Court," the quasi-monarchial circle established around President George Washington. The "Court" label became associated with the home of Anne Willing Bingham, who had returned to Philadelphia in 1786 after three years traveling in Britain and on the Continent.[10] The manner in which "the women of France interfere in the politics of the Country, and often give a decided Turn to the Fate of Empires" inspired Bingham to establish her salon. These entertainments, which featured aristocratic displays of luxury, risqué conversation, and an exclusive list of invitees, scandalized republicans and even some Federalists. As a *salonnière* she was joined by Elizabeth Graeme, also a European traveler. Graeme's salon was more sedate than Bingham's, less politically charged and salacious. Yet both set a tone for Philadelphia social life that lingered far into the nineteenth century. Salons, balls, and dinner parties were not retreats from the political sphere of strife and contention but public sites in their own right. The ancien régime tone of Bingham's affairs, in particular, was resurrected in the parlors of the Carolina Row.[11]

The reactionary sociability that came to be practiced on Spruce Street also had its roots in the Old World. Ralph Izard, who had one of the largest slave holdings in South Carolina, had been educated in England before returning to Charleston in 1764. He drew an income of about £5,000 a year from the labor of his slaves, who were managed by an employee since Izard claimed that "the Climate of Carolina will not suffer me to take the management of my estate into my own hands." During one of his trips to New York he met Alice DeLancey. John Adams described her as "a Lady of great beauty and accomplishments as well as perfect purity of conduct."[12] The Izards embarked for England in 1767, traveling widely there and on the Continent. He returned to America in 1781; Alice did not return to South Carolina until 1784. She had spent nearly a third of her life abroad.[13] Between 1768 and 1792 Alice Izard gave birth to fourteen children, only seven of whom reached adulthood.[14] Five of the surviving seven spent their formative years in Europe. When she set foot once again on American soil in 1784, Margaret (b. 1768) had spent nearly all her life abroad. The family became enamored with the high culture of prerevolutionary France. Yet they also possessed a strong sense off nationalism. As Margaret's friend Josephine du Pont observed, "in loving France and the French you knew not only how to preserve your character and native tint independent of foreign impulses, but also keep the superiority they give you in foreign society. I imagine it would be almost insulting to say of you 'She is completely French' even taking it in the best sense." The Izard women were European, but they were emphatically American as well.[15]

On May 1, 1785, seventeen-year-old Margaret wed Gabriel Manigault. He too had traveled widely in Britain and on the Continent. He had studied law at Lincoln's Inn and fancied himself an architect, but he had no stable profession. He hardly needed one. As one of three heirs of his eponymous grandfather, reputedly one of the wealthiest men on the mainland, he could live as a gentleman of leisure. He made an honorable attempt to manage his family's slaveholdings, but to no avail. Margaret reported that such duties were "precisely the thing of all others which he detests. He cannot endure that sort of business, & finds neither credit nor profit by it." Between 1791 and 1806, Margaret gave birth to ten children. Only three—Charles (d. 1874), Gabriel Henry (d. 1834), and Harriet (d. 1824)—survived her. Yet during the years the family resided in Philadelphia, the household was overwhelmingly feminized. Five

Manigault girls—Elizabeth (1792–1822), Charlotte (1792–1820), Harriet (b. 1793), Emma (1797–1815), and Carolina (1803–1818)—occupied the lion's share of their mother's and grandmother's attention, as the boys were often in Europe or away at school.[16]

Neither Gabriel nor Margaret Manigault was happy in South Carolina. Gabriel disliked the life of a plantation patriarch, and neither could abide the lowcountry climate. Both she and her husband pined for a more cosmopolitan society than Charleston, with all its charms, could provide. The family relocated to New York, where they had relatives, in 1805. The children adjusted well, particularly enjoying the "snow & ice," but their parents were less satisfied. Knickerbocker society was hardly more cosmopolitan than Charleston's. The summer heat, radiating off New York's "paved streets & rowes of brick or stone buildings, all touching each other with no interval for a circulation of air," proved nearly as stifling as the climate of Charleston. Worse, six of their seven slaves soon escaped ("to our great disgust & relief," Charles recalled in a Reconstruction-tinged pique). Philadelphia was an attractive alternative. Ralph Izard and Gabriel Manigault had traveled there often and had numerous friends and business associates there, particularly John Vaughan, an official of the American Philosophical Society whose family they had known in England. In 1807 Gabriel purchased a large estate on the Delaware River about twenty miles above Philadelphia. China Retreat (rechristened Clifton) was a substantial investment even by the standards of the Manigaults and Izards. But it was a worthwhile one; the Philadelphia area became the center of family activity for the next twenty-five years.[17]

The site was ideal. Elevated and well wooded, the house was cool in the summer. It was adjacent to Farley, George Izard's estate, and Philadelphia, where Margaret had many close friends. It was distant enough to isolate the family from urban outbreaks of yellow fever and cholera. It also possessed the scale and magnificence of a true country house. Charles Manigault recalled it as *"a Magnificent Dwelling. . . . Beautifully Situated"* amid woods, *"cultivated lands,"* and "good neighbors." Joshua Francis Fisher, a Philadelphia gentleman who was close to Charles and the young Manigaults, remembered Clifton fondly as "a very large house, with rooms . . . of palatial proportions."[18] It soon became their permanent residence. The Manigaults took to renting "very well situated" houses in the city so to enjoy the busy winter social season. By 1808 the Manigaults were consistently listed in Philadelphia directories as living in various addresses in its downtown district.[19]

In 1797 Ralph Izard suffered a debilitating stroke. Until his death in 1804 his wife cared for him either at their house in Charleston's South Bay neighborhood or at the Elms, the Izard family's plantation on the Cooper River, outside Charleston. In 1809 catastrophe struck again: Gabriel Manigault, aged fifty-one, died suddenly at Philadelphia of an "awful & subdueing affliction" of "paralysis"—most likely a massive stroke. The death was a shock to the family, but particularly to Margaret Manigault. By all accounts theirs was an affectionate, even egalitarian, match. Confronting the debts accumulated by her absentee planter husband and considering her responsibilities as a single parent to seven children aged six to seventeen, she decided to rent out Clifton in order to buy a town house in Philadelphia. Maintaining an unproductive country house was simply out of the question. In the city the children could be educated and introduced into society at appropriate ages and the family could live off the income from its Carolina estates.[20]

In 1816 she entered into negotiations with Luis de Onis, the Spanish chargé d'affaires, to acquire his house at the northwest corner of Ninth and Spruce streets. Boasting a Mansard roof allowing for living quarters to be placed in the attic, the house contained all the "ingredients in the composition of Comfort & well-being"—large, airy rooms; a large lot with a garden; a central location in the city's fashionable district; and available lots on the block where other members of the family could relocate. At $20,000, the cost was prohibitive. But as Margaret mused to her son, "If ever I am to own a comfortable house in Town, it is time to get about it. A mon age, les années tout precieuses." Until her death in 1824 her house and the Carolina Row it anchored were the center of a transatlantic social circle dedicated to preserving the conservative values of continental Europe in a liberalizing nation.[21]

"They are the reformers of the World"

After recovering from the deaths of their husbands, the widows faced two main dilemmas: the education of the Manigault children and the management of their introduction into society. It was essential that the boys be given access to professions that would allow them to manage the family's fortune and to prepare them for public service. As Gabriel Manigault explained to Harry, his goal should be to become an "independent gentleman . . . that station you will certainly find the happiest in the world."[22] Harry and Charles were steered toward mercantile and military careers. When the latter left the army in 1815, his mother admonished him to stop "lounging around" Philadelphia and investigate its resources in *"law, medicine, commerce, & others."* Ironically, Charles became a planter who took great pride in his hands-on management and paternalistic ethos.[23] Charles, and most likely Henry as well, had dancing, fencing, violin, and French masters while at school; the whole family had a solid grounding in French, it being the language of choice *en famille.*[24] Boys were also expected to excel at penmanship and letter writing. In 1812, while Charles was away at school, his mother chastised him for being a poor correspondent. "Attend to your spelling," she ordered. "It is disgraceful for a gentleman to be deficient on that point." And, of course, gentlemen-to-be also had to master the necessary social graces. In many respects, then, their education overlapped with their sisters'. But Alice Izard and Margaret Manigault took special care to see to the education of the family's young women.[25]

In some ways the stakes were much higher. True, young men had to be groomed for the "world"—the battlefield, countinghouse, and legislative hall. And, of course, they had to provide for their families. Women also had a public function, as the widows illustrated vividly in the operation of their salons. But young women had a smaller margin for error than boys. Harry, in particular, engaged in behavior that would have rendered any of his sisters spinsters for life. He was a "wild, uncontrollable youth," Charles recalled. Yet in spite of this debauched reputation, he married well and settled into the life of a planter in 1825. It was not that the adults did not try to straighten out young men. To the contrary parents nagged sons persistently. "What is it for a father to write to his son?" asked Gabriel late in 1808. "Always the same thing, nothing but advice, advice, advice." And not just advice, but correction also; in 1814 Margaret chastised a Middleton relation for an unspecified but serious "impropriety" and admonished another to abandon his misanthropic wish to "absent himself entirely from society."[26]

The girls could afford to be tainted neither with dissolute nor antisocial behavior. Scandals made the rounds swiftly, and the more delicious the better. In 1816 Henry's mother related a "frightful" rumor about a Philadelphia lady who "had left her poor husband in England, & followed an officer of the guards to the Continent." Their salacious nature was one reason such rumors were so popular, but they were also useful tools for instruction. "What would have become of" the adventurous wife if the rumor were true (she doubted it), Margaret asked? The hard fact was that her girls' welfare—to say nothing of their happiness—depended on their choice of a husband. Their reputations had to be pristine. It was to their education, then, that Alice Izard and Margaret Manigault devoted themselves.[27] None of the Manigault girls went away to school. They were educated at home by their mother and grandmother. Both women shared the concerns of educational reformers such as Benjamin Rush (the Manigaults' family physician) and Judith Sargent Murray. These reformers—who argued that mothers, wives, and hostesses exerted real influence on the moral culture of the nation—insisted that women needed to study subjects such as history, the sciences, and rhetoric.

The Carolinians were far more conservative, however. Their sought not to establish a new model for educating republican young ladies but to adapt an old one: that of prerevolutionary France. They had little patience for institutions of any kind. "Boarding Schools," proclaimed Alice Izard from New York in 1808, "are not proper places for the education of young ladies." Fashionable seminaries fitted women neither for the nursery nor for the parlor, but only to be "always in the streets, galanted by Gentlemen." If American society needed an infusion of elitist principles, young women had to learn the morals and manners befitting a republican aristocracy. Alas, "the fashionable ladies of the present day have not been educated to these ways." It was up to Izard and her daughter to rescue elitism from fashion and frivolity. The education of their girls would be an extension of a larger political struggle against the extremes of democracy. "Our well bred gentlemen must set the example, & so must our well-educated Ladies," she insisted in 1811. "They are the reformers of the World."[28]

The young Manigault women received an education befitting their social status. They studied literature and language, music and dance, and some formal academics. They spent no time at all on vegetable gardening, cooking, sewing, cleaning, and the like, for these tasks were for servants, slaves, and hired help. As Mary H. Middleton observed during a trip through a yeoman section of South Carolina, "in *these parts* all that [is] necessary for a woman to know is the curing of bacon & making soap. You will allow that those accomplishments are incompatible with studying Montaigne." Her glibness notwithstanding, education for girls was seen as a serious matter. The young ladies of their circle would never need to know how to smoke a ham, but they would embarrass themselves if they played the harp clumsily or fumbled a droll comment because they lacked fluency in French.[29]

Compared with the attention the young women of the family lavished on literary pursuits, musical accomplishments, and physical fitness, they received surprisingly little instruction in the principles of religion. They usually attended services on Sundays, though they did not attend any particular church or denomination with any regularity. In fact, they quite unconventionally made occasional attendance at the city's Unitarian church. They had little sympathy for strong expressions of piety, but viewed religion as the foundation of sound morals and social order. Alice Izard saw little to admire in Unitarianism, but she approved of her daughter's attendance at services with

John Vaughan because she "like[d] him, & his benevolent ways." Extreme enthusiasm, like impiety, was to be distrusted. In early 1809 the family was introduced to Nicholas Biddle, who had just returned from a European tour. He regaled his hosts with an account of the phrenological theories of Franz-Josef Gall. The German argued that "every sensation has its proper apartment" in the mind, and by "knowledge of the arrangement of the brain" one could suppress vice. How vague are such theories!" Alice marveled. "How little capable of leading to happiness."[30]

Religion allowed gentlepeople to enjoy life's pleasures without surrendering to cynicism on the one hand or sensuality on the other. Thomas Loughton Smith had succumbed to the former. This South Carolina gentleman was "too much practiced in the ways of the world, to *feel* again, & that is a cheerless prospect for him," Alice reflected. Dissipation, by contrast, had been the fate of Julie de Lespinasse, the Parisian *salonnière.* In 1810 Izard was looking out for a copy of the Frenchwoman's recently published letters, though Alice knew "what I have to expect & lament that so much sense, & so much talent should have taken so violently wrong a bias." The Frenchwoman's life illustrated "What poor creatures we are without the guidance of our best friend religion. Morals soon fall a sacrifice, & are unable to sustain themselves against the war of the passions, & the exertions of feeble reason." Religious faith should produce not personal transformation but resignation to one's place. It was easy for privileged women to resign themselves to their station, of course. Yet religion did offer comforts to a mother and daughter who buried their husbands and many of their children during their lives.[31]

Religious instruction was limited to girls, but both sexes were expected to keep physically fit. Boys enjoyed the liberty to exert themselves in sports and rough play. Genteel expectations created more problems for young women. Excessive physical activity was deemed unfeminine, but enervation and inactivity were equally undesirable.[32] Thus, the girls took walks in the country or in safe areas of the city. In 1808, however, their grandmother hit on a novel and newly popular alternative: horseback riding. Her preferred method of exercise, dance lessons, was (temporarily) out of her favor because current fashions seemed less like dance than "violent gestures which deserve the appellation of bucking." Riding, though, promised to be "more pleasing from its novelty & more conductive to health." Izard eventually settled on a school run by a Frenchman popular with Philadelphia's belles. His methods enhanced both vitality and refinement in near-perfect harmony. "I think the exercise will improve their health," Alice told the girls' somewhat skeptical mother.

"Their manner of holding themselves on horseback will be of use to their carriage and make them hold up their heads even better than a dancing master."[33]

Though the Manigault girls devoted some time to religion and physical exertion, they were far more engrossed by formal academics, dance, and musical instruction. Knowledge of botany and French literature was inseparable from mastery of the cotillion, the harp, and (if one had the talent) vocal music. It was not merely that their possession separated the refined from the vulgar, though that was important. The Manigaults' circle believed that literature, science, music, and dance upheld Euro-American civilization, and that it was their duty to help maintain it. It was also essential knowledge for a hostess. One just could not entertain effectively without knowing how to dance, to steer a conversation, or to animate a company with a song or a performance. Academics, refinement, and sociability were discrete goals, but shared one overarching aim—to turn girls into gentlewomen who would wage the fight for elitism in the ballroom and salon.

In the sense that study, music, and dance were part of the fabric of the Manigaults' lives, it is artificial to isolate them as part of a discrete educational regimen. Still, knowledge and taste had to be acquired, skills to be learned. This was a remarkably literate family. As reactionaries, the family confronted new cultural trends with skepticism, if not outright distaste. The children's reading thus reflected their elders' opinion that the "authors of the passed [eighteenth] century were wise, & clever beings & what is worse for us, I do not think the present one has improved upon them." The girls' study, not surprisingly, rarely strayed from well-worn paths. Modern learning was not just a waste of time, it was absolutely poisonous. Botanical researchers, having exposed nature "in even her most apparently delightful productions, plants, and flowers . . . have rendered what was once an innocent, & delightful amusement, almost a reprehensible one," Alice Izard averred. She was hardly anti-intellectual, however. Rather—as her comment on phrenology indicated—she objected to the creeping secularization of western cultural life. Piety should be the foundation, not the focus, of intellectual endeavor. Its absence bred the guillotine, Napoleon, and cultural declension. "I sometimes think that it is owing to the young people of our times attending to such discussions that they are so bold, & indelicate," she concluded warily.[34]

Much that the youngsters read was thus "safe" material: Gibbon, *The Spectator,* scripture, de Sévigné. Harriet Manigault described a typical day of lessons in 1814, which began with breakfast at nine A.M. An hour casually reading French (Marie-Thérèse Lamballe's *Mémoires*) while pacing the music room followed a more rigorous period of translating French to English (La Rochefoucauld's *Maxims* and a book of ancient history),[35] then a break for half hour or so on the harp or piano. Afterward the children clambered up to their mother's room to listen to some Bible passages, followed by some lessons from Pricilla Wakefield's popular book on botany. "Mrs. W. is not very deep," Harriet observed, perhaps in praise or perhaps in complaint.[36] Then Emma Manigault read aloud from *Decline and Fall of the Roman Empire* for thirty minutes; her two sisters followed with Chateaubriand's travels through Greece and the Holy Land and Barthélemy's account of ancient Greece.[37] These formal lessons ended around two in the afternoon, after which the girls "usually employ ourselves as please." The dinner bell rang at three, followed by more music, a carriage ride, and perhaps some social calls. Somebody read at afternoon tea and evening knitting, "generally some *novel,* as we read such serious books in the morning." In the summer of 1814 that book was Elizabeth Hamilton's *Memoirs of Modern Philosophers,* whose Burkian morals and familiar family arrangements (three young women with a group of older, independent women in the background) must have resonated with this conservative family. After supper between nine and ten there was conversation, cards, and chess—and then to bed at eleven.[38]

Though much the girls read was unchallenging, their elders did not shelter them from dangerous new ideas. Young people needed to know what was going on, if only to avoid being ambushed by it. It was important, in a house frequented by active members of the American Philosophical Society, that they be able to converse intelligently on scientific subjects. Likewise, it was necessary to keep abreast of literary developments, however repugnant they were. In 1812 Margaret Manigault hungered after the recently published letters of Anna Seward, a copy of which she acquired from John Vaughan.[39] She harbored no sympathy for the Englishwoman's "abominable" style, and Seward's political views (admiration of Mary Wollstonecraft, her condemnation of the "remorseless Washington" in *Monody on the Unfortunate Major André* [1781]) won her no friends on Spruce Street. Nevertheless, Seward was popular, and sentimentality was on the ascendancy. Manigault explained, "I have got used to her frightful & uncouth turns." She also expected her children to keep up with literary developments. In late 1812 she urged Elizabeth Morris, her eldest daughter, to subscribe to the *Select Reviews of Literature, and the Spirit of the Foreign Magazines.* "You would find sentiment & instruction in them," she explained. "They tell you what is going on in the *Literary* world." Margaret regarded engagement with the world as a class responsibility. Cultural rot could not be reversed with ignorance. Gentlefolk like the Manigaults distinguished between learning and bookishness. Culture provided entertainment and uplift, but was never an excuse for antisocial behavior. As she assured her daughter, "I am not afraid of your growing a *Blue Stocking.*"[40]

Just as fluency in French and familiarity with the zeitgeist were serious endeavors, not diversions, they also pursued dance and musical instruction with a purpose. Music provided entertainment in quiet, private moments, but, more important, it made for a good party. The Manigaults spent freely on lessons, scores, and instruments. In 1800 Margaret boasted to a friend that her "drawing room is now disfigured by a grand piano, whose delicious tone makes us forget its deformity." All the children received instruction in a variety of instruments. The boys played the piano, violin, and guitar. In 1808 Alice Izard hired Anne Marie Sigoigne, a refugee from St. Domingue to whom they had been introduced by Nicholas Biddle, to teach her granddaughters harp and piano. She was not a talented musician, Izard conceded, "altho' she is said to be a good teacher & they think her so." Her daughter Adèle was, however, a "marvelous" harpist (she posed with a harp in Sully's 1829 portrait). Their instruction was intensive; the Sigoignes taught the girls for several hours every other day. To complement their lessons the girls attended concerts and recitals of church music at Philadelphia's Catholic and German churches.[41]

Dancing had physical as well as social benefits, since it developed graceful movements and correct posture. But like literature, fashion, and language, dance was a contested issue in the culture wars of the early American republic. The issue confronting the Izard and Manigault clan in the early nineteenth century was the smash popularity of the waltz. It differed from traditional court dances such as cotillions and quadrilles because of its vivaciousness and the close contact it necessitated between women and men. The Manigault girls learned traditional dances as a matter of course. Dancing was not only fun and sociable; it also developed appropriate manners and morals. Alice Izard praised the dancing of the children of a Russian diplomat, observing "Their attitudes are all ease, & their ease is not diminished by the precision of their steps. They laugh, & talk, & jump, & play, & dress modestly & becomingly." Gentility was hardly effortless, but it was essential that it appear so. Likewise, traditional dance embedded discipline at the heart of pleasure, inculcating habits of restraint while giving license to physical enjoyment and sociability.[42]

Not so with waltzing, at least as it became practiced in the early 1800s. Alice Izard frowned on new dance steps in general although she was self-aware enough to ascribe her attitudes to generational prejudices. So when in 1809 she took her grandchildren to a popular dance instructor and found no merit in "the new steps, the new attitudes, nor the present manner of dress," she still allowed the girls to learn them. Waltzing was different. Even though many critics panned it as immoral from its inception, it became popular in pleasure-loving aristocratic circles toward the end of the eighteenth century. Early in the 1800s a new style of waltzing became popular. Where earlier steps were lively, the new ones seemed frenetic; where women and men had previously held each other closely, their bodies and even faces now touched. The root of the problem, of course, was revolutionary France. Mary S. Pinckney, accompanying her husband on his diplomatic mission in 1796, attended a ball in Amsterdam where the waltz was danced "with a great deal of decency." But when she learned "it was not danced with half the spirit . . . the French ladies in general are capable of throwing into it," she refused to allow her daughter to join in. The corrupted version had crossed American borders by 1800. At an 1805 party Margaret Manigault observed dancing that "was not the sedate & decent walse which we used to see at" French soirees in the 1770s. She endorsed her brother's assessment that republican France "ha[d] altered the nature and corrupted it." The problem lay not with waltzing's European or aristocratic roots. Rather, Manigault resented the appropriation of a genteel pastime by the mob, a specific manifestation of a broader movement: the intrusion of ordinary people into areas previously controlled by a social and political elite.[43]

". . . agreeable amusement"

Education was pursued for its own sake, for personal fulfillment. But the social functions of education transcended its individual benefits. It taught children—girls in particular—how to behave in society: how to shine in genteel companies, to attract desirable husbands, and to entertain effectively as hostesses in their own right. Polite society was, at its root, anything but polite. Intense competition for status lay at the heart of the genteel world, the greater given the absence of formal titles in a republican polity. Gentlefolk often protested to the contrary, but wealth was the most important determinant of status. Still, it was hardly the only one. Family name, travel (especially travel to Europe), education, manners, and mastery of genteel skills such as dance and dress were all variables in the complex equation that generated social status in early republican America. Thus, an individual's behavior in society was crucial in acquiring and maintaining rank. But competition was not the only feature of these affairs. In an era when critics condemned genteel society as immoral and un-American, the gentry struggled to maintain the integrity of its culture. Though less visible than grand entertainments, everyday rituals of politeness such as social visits and small-scale gatherings also helped privileged women and men hash out status and maintain genteel standards.[44]

Social calls entailed dropping by the homes of one's close friends and family members for conversation and perhaps tea or a light meal in the morning or late afternoon. This was most easily done in cities, though when the Manigaults lived at Clifton they often rode out into the countryside to visit Andalusia, the Woodlands, or another estate. Hospitality at such houses affirmed one's social position. When she received an invitation for afternoon tea from the daughters of the Swedish minister, it gratified Alice Izard "by showing how much they valued my Daughters & Grand daughters." Women and men both shared the obligation to give and receive visits. When in 1808 Harry Manigault tried to beg off his duties, his father admonished him that social calling was "a matter of real consequence in society, & one on which a young man's character in the world often depends." Women devoted themselves to the rites of private visiting. It enabled them to keep up with current styles in dress, hairstyles, and other fashions. Visiting also facilitated the exchange of news and gossip. The imperative to see and to be seen was so powerful that even Alice Izard, so cozily ensconced at her son's estate that she had "not a wish to go into any other house," felt compelled to "conform to the customs, & manners of the family." To remain home would "oblige them to break thro' rules of society [and] neighborhood, which . . . might be attended with disagreeable consequences." So she chatted with the Craig family, and accepted their invitation to dine, which was not such an unpleasant duty, after all.[45]

Harriet Manigault turned the polite visit into a kind a social scoreboard. Having heard so much about a Miss Keene, a Philadelphia lady, the family's Carolina friends demanded an introduction during an 1815 visit. But as Harriet wrote to her sister, the fashionable Miss Keene failed to live up to her reputation. She "is very much altered, she has quite a *haggard* look," Manigault reported cruelly. The manners of all the ladies in attendance were very much on display, every habit and article of dress fodder for judgment. Keene's appearance disappointed, but so did her manners. Compared to the "modest & unassuming" behavior of one cousin, "so much like a lady in every respect," Keene seemed "*made up* & conceited, every word that she utters gives one the idea of having been studied." Fashionable women were very much aware of the theatricality of even so prosaic an affair as the social call. As Manigault said of Miss Keene's friend Mrs. Lenox—another close friend of her mother's—her "ancient charms were exposed to the gaze of the public, just as if they were something very delectable to feast one's eyes upon."[46]

Travelers especially benefited from the routine of visiting. Exchanging social visits provided news and reintegrated them into the routines of genteel life. And in a well-traveled circle such as the Carolina Row's, visiting had an important cultural dimension. As travelers compared fashions, discussed political and cultural developments, and introduced newcomers, they developed a national standard of cultivation that transcended, without obliterating, local peculiarities. In 1797 Eleanor Parke Lewis of Virginia reopened a sealed letter to her friend Elizabeth Bordley to tell her that Margaret Manigault had just visited her at Mount Vernon on her way from Philadelphia to South Carolina. Manigault "charged me most sweetly to remember her affectionately to you," Lewis wrote. The Virginian returned the favor years later when she visited Philadelphia and promptly engaged in "the *agreeable amusement* of morning visits." Without the information supplied by social calls, travelers might commit a faux pas when they entered a more general company.[47]

Another "agreeable amusement" the family orchestrated was what Margaret Manigault called "our society of old gentlemen." This was a small, private company that enjoyed a standing invitation to gather at the houses of the Carolina Row in the evenings for dinner and conversation. Like social calls, it was both cooperative and competitive. There was little reason to attend closely to display and form at these private companies. Alice Izard and Margaret Manigault favored it precisely because its tone represented a refuge of sorts from the dizzying social whirl in which they participated. But a powerful element of competition also lay at the heart of these companies. The widows and some of their elderly guests disagreed sharply over political issues, so verbal sparring was a distinguishing feature of these affairs. They never threatened to devolve into incivility, however, because the participants respected the rules of polite society, which provided pleasure, on the one hand, and protected civil discourse, on the other.[48]

John Vaughan, William Short, and José Corrêa da Serra were the core of this intimate society. It met throughout the Manigaults' years in Philadelphia with special intensity during the war years of 1812–15, when her boys Charles and Harry were away in uniform. John Vaughan was the family's oldest and most ardent Philadelphia friend. Born in England in 1756 to a prominent Unitarian family, he immigrated to Philadelphia in 1782. By 1791 he had established himself as a wine merchant and business manager for a number of South Carolina families, including the Izards. For nearly the first four decades of the nineteenth century he served as the librarian and then the treasurer of the American Philosophical Society, in whose apartments he resided. Vaughan was a Federalist, though not a dogmatic one. He was better known for his benevolence and sociability; a friend later recalled that Vaughan possessed a "practical goodness" that manifested itself in his assistance toward travelers of note arriving in Philadelphia. Vaughan was "particularly intimate with the colony of Carolinians on Spruce Street; he claimed all their friends as his own."[49]

William Short (1759–1849) had served as Thomas Jefferson's private secretary during his ministry to France in the 1780s. Initially an enthusiast of the French Revolution, he turned against it in horror with the onset of the Terror in 1792. He served in a number of diplomatic posts on the Continent, but when the Senate declined to name him as the first minister to Russia—and after his romance with Rosalie de la Rochefoucauld ended with her marriage to another man—he returned to the United States for good in 1810. He settled in Philadelphia, where he made a fortune as a land speculator. The Abbé Corrêa da Serra, the renowned Portuguese botanist who became minister plenipotentiary to the United States in 1816, was the most controversial member of the family's circle of old men. The abbé lived in the United States from 1813 to 1820, spending the bulk of that time in Philadelphia. He had known both Short and Nicholas Biddle in France. Their acquaintance, in addition to his fame as a wit and conversationalist, won him access to Philadelphia's most fashionable homes as well membership in the American Philosophical Society and its social circles. Corrêa's early enthusiasm for the French Revolution was tempered by Napoleon's dictatorship. Though chastened, he remained committed to republican principles throughout his life.[50]

These three men shared qualities that made them treasured companions to the Spruce Street widows. Most important, they possessed an eighteenth-century temperament. They provided companionship of the kind that Alice Izard and her daughter had not enjoyed since the deaths of their husbands. Little surprise then, that the family gathered their "little *Societé* almost every evening," as they reported during the cold winter of 1813. As living embodiments of eighteenth-century standards of comportment, these three men might serve as models for the Manigault children. Not only was a "sensible, well-informed, polite old man" a "great resource in the long winter evenings," Margaret explained, but "a treasure, I think, in society, & particularly in a large female family like ours." Vaughan, Short, and Corrêa instinctively knew how to balance politeness with argument, wit with seriousness; their society was anything but banal. And, of course, their cosmopolitan backgrounds—together, the widows and their guests had spent more of their lives abroad than in the United States—made them congenial companions. Izard and her daughter exploited these three men in their crusade to advance quasi-aristocratic values in the hostile atmosphere of the early republic, but they knew that time was short. "Well educated men of sense grow more and more rare," Margaret confided to her mother in 1812. "I very much fear the breed will be lost."[51]

The Carolinians were committed to revitalizing elitism for a new century. That necessitated engaging with distasteful new ideas. The society of these old gentlemen also served that end since they possessed such a wealth of experience and diverse views on the affairs of the day. "Good Vaughan," though, avoided controversy. He served mainly to brighten the family's spirits with his buoyant personality. Short complimented the family by his attachment to French culture. "He is French in his soul," Margaret observed. "He does not understand how one can imagine living, or possessing what we emphatically call *Comfort,* outside of France." Short crossed verbal swords with the republicans in the family's circle, particularly the abbé. Margaret rejoiced on hearing in late 1812 that Corrêa was "*very impatient* to be invited" to the Carolina Row, for his conversational skills were legendary. He was a fit adversary for the reactionary gentlewomen of the household. "The Abbe reasons, & moralizes with me while my Mother & three others play at Whist," she reported happily early the next year. Exposure to Corrêa's heterodox notions was as important as familiarity with other dangerous ideas. It insulated young women from an increasingly sentimental popular culture, fashioned them into companions, not merely mates, for prospective husbands, and armed them for battle against the forces of democracy. "The society of sensible men improves the female mind," declared Margaret. "There is a degree of solidity & of elasticity even upon trifling subjects with well educated men—which compared with the vapid stuff that occupies most of our ladies is very overpowering."[52]

"I really need a little dissipation"

Intimate social affairs could satisfy neither the family's sociable instincts nor its ideological goals. Though the adults often sang the praises of quiet hours spent within the family circle, in fact they loved balls, dinner parties, and other grand affairs. These select companies allowed the Manigaults and other prominent families to affirm their high rank, demonstrate their wealth, and distinguish themselves from social inferiors. These posh entertainments were also fun, of course: the enjoyment of music, dancing, and sumptuous foods into the early hours were privileges of a leisure class unburdened by distractions like jobs or bill collectors. Political considerations also prompted the Manigaults to enter society. The democratic drift of the new republic alarmed them profoundly. If it was allowed to continue, they feared that all distinctions would disappear. The public role assumed by cultivated women seemed threatened by a culture that defined equality by gender instead of class. Praising de Staël's *De l'Allemagne* (1810) as the work of a "sublime genius," Margaret Manigault complained that reviewers "do not wish to believe that a Woman can guide an Eagle's pen." As women who took their public roles for granted, the Carolinians resolved to take action. Alice Izard admonished her daughter, "your well-regulated family & your amiable daughters might do a great deal" to reverse the tide of American culture. "Make your house again pleasant to acquaintances as well as friends & particularly to Strangers of character. This you have done," she urged, "& you must not give it up! It is productive of too much good to be laid aside." In other words she urged her daughter to establish a salon.[53]

Somewhat reluctantly, given her responsibility for a large, young, and frequently sick family, Margaret complied. Her salon met by invitation only about once a week during the busy winter social season and sporadically at other times. Its inner circle consisted of her resident family and any related Carolinians in town, close family friends from Philadelphia such as Elizabeth Bordley, Robert Walsh, the Hopkinson, Craig, and Biddle families (connected thereby to Joseph Dennie's circle), the "supper tray" crowd of Vaughan, Short, and Corrêa, and several families from the resident European diplomatic corps.[54] It was, given its reactionary mission, an overwhelmingly Federalist company. These affairs differed from social calls in a number of ways. The imperative to perform was less significant in these affairs. They offered a much wider variety of activities than social calls and the old men's circle, which mostly consisted of conversation. The salon certainly featured talk, but it also offered diversions such as whist, musical performances, games, and dances. These intimate gatherings illustrate the cohesive power of sociability. Shut off in their dining room and parlor, the Manigaults and their guests could behave as they pleased without exposing themselves either to the prying eyes of fashionables or to the derisive remarks of ordinary folk.

Neither its Federalism nor the leadership of women distinguished these salons from other contemporary companies. Its ancien régime tone, though, certainly was unique. "Only in France can one enjoy life," Margaret Manigault declared in 1814, inspired by news of Napoleon's exile to Elba. But relocating to France was not an option; for these nationalists found in the United States "some charm which is irresistible. . . . a security, a simplicity, a truth, a freedom from care." Besides, in France the children "would perhaps love their country less." So they settled for fashioning a little Paris on the Delaware.[55] Befitting the Carolinians' admiration for "La Belle France," three families of the resident diplomatic corps enjoyed standing invitations to the salon: those of Luis de Onis, the Spanish minister; Baron Johan Albert de Kantzow, minister from Sweden and Norway; and Andrei Dashkov, Russia's minister.[56] These families were culturally and politically conservative, but (in the cases of the Kantzows and Onises) they also had children of the Manigaults' age, and (in the cases of Kantzow and Dashkov) their wives radiated wit and elegance. The young daughter of Luis de Onis possessed a remarkable singing voice. William Short expressed "astonishment" at her skills after hearing her at an 1814 recital. The two Kantzow girls, welcome always to "come in casually and without invitation to take tea," were "gay, kind, simple, enthusiastic, a little noisy," and "rather large for the taste of Americans."[57]

Young people provided merriment, but adults provided gravity and companionship. Margaret Manigault confessed she was "dying to be in company with" Mme. Kantzow when Robert Walsh whispered that the Russian was "without exception the most *accomplished Lady* that was ever seen in America." The two became fast friends, and the Kantzows became regulars at the Manigaults' salon. Margaret and Andrei Dashkov's wife hit it off soon after her arrival in Philadelphia in 1809. The two were bound together by their loathing for postrevolutionary France. In 1813 word reached Philadelphia that Louis Serrurier, Napoleon's envoy, had become "enraged" when Dolly Madison walked arm-in-arm with Dashkov around a White House gathering. "This kind of silliness occupies and amuses that Court," Manigault beamed. "I admit it delights me." Dashkov too became a regular attendee at the Spruce Street affairs.[58]

These Europeans felt far more at ease with the Manigaults than they did in most American companies. Margaret Bayard Smith, a leader of Washington society, complained that the Onises and Kantzow girls "always sit together, stand together and talk together and never join any of the other young ladies."[59] The source of the Europeans' discomfort is not hard to find. Like the socialites satirized in James Fenimore Cooper's *Home as Found* (1838), the American gentry looked on Europeans simultaneously with awe and disdain (though Cooper emphasized the former). They represented the genuine aristocratic article, but republican ideology disparaged formal rank as an anachronism. As a result, Americans behaved awkwardly around visiting Europeans, which made their guests uncomfortable in turn. The Manigaults and Izards suffered from no such inhibitions. They did not strive to recreate Parisian society, an unreachable (and undesirable) goal. Rather their salons blended American and European forms. As Margaret explained, if "well-bred foreigners . . . see that you wish to please them, & to place them at their ease, they are grateful, & enjoy the happy privilege." She knew the resident diplomats did not care about the minutiae of manners and consumer items that preoccupied so many of those entranced by the ideal of refinement. They sought out Americans who were comfortable in their own skins.[60]

In these small, intimate gatherings the Manigaults and their company indulged in entertainments that invited censure by moralists. The Spruce Street women offered their guests a variety of entertainments, particularly games, music and dance, and conversation. Backgammon, chess, and checkers were innocent enough. But card games such as whist and piquet were also very popular among the set that congregated on the Carolina Row. Critics of polite society condemned games of chance as immoral. Even some gentlefolk had qualms about the practice. Eliza Lucas Pinckney tut-tutted at British card players during her 1753 visit. The Carolinians were unbothered by such practices. After one soiree in early 1814, Margaret Manigault expressed bewilderment at opponents of card playing. Some of the company gathered around the piano, others around the fireplace for conversation. "We wanted a card table to occupy some of the désoeuvrés." Referring to critics of such practices, she wondered, "Is it not surprising that the merits of a Card table should not be understood?" What, she implied, could be more offensive than boring one's guests?[61]

Invariably a young crowd congregated around the piano. The family ignored nationalist pressures to patronize American composers. The girls and their guests took turns entertaining the audience on the piano and harp, alternating between instrumental and vocal music. Perhaps because listening could not satisfy the young guests, the audience usually organized an impromptu dance. While balls and assemblies, with their large crowds and competition, encouraged precise steps and attention to decorum, in the privacy of their own parlors the Manigaults and their guests felt free to indulge themselves. At one private 1814 gathering in Clifton's spacious ballroom, the company "waltzed & danced & jumped & *squealed* & sang & laughed & talked." The adults would never have sanctioned such a lack of restraint in a more public setting. The Manigaults and their circle believed in enjoying, in Charles's words, "the rational & social pleasures of this world." What was the point, after all, of having all that money?[62]

Salons also featured conversation, of course, and Margaret Manigault's reputation as an "esprit fort" guaranteed that guests could expect substantive talk on cultural and political affairs. Joshua Francis Fisher, whose mother was a close friend of Margaret's sister Nancy Deas, recalled that because Manigault was "singularly agreeable in conversation," her salon became "the resort of all the intellectual and refined society of" Philadelphia. She kept an eye out for interesting, accomplished women and men. Thus, she was disappointed at one Mrs. Derby who, while "entertaining when she speaks of what she has seen in her travels," nevertheless betrayed "a affectation for literature, a taste for the fine arts, [and] of speaking upon subjects which she does not understand." Although it was important, conversation constituted just one aspect of these affairs. It was to merge with music, dance, and games to recreate "the ease and elegance of French manners" in an American context. Because these affairs were select companies, they more easily combined what Charles called "the formality & etiquette in the elite of society" with insouciance. Joshua Fisher remembered that at Alice Izard's Monday-evening receptions, she displayed "refined manners" and "graceful dignity." But her parties were hardly stiff and joyless. "Many a pleasant evening have I passed there," Fisher recalled. The combination of propriety and ease combined "to give a charm to these simple receptions, which the utmost luxury could not have enhanced." At exclusive companies guests could be certain of their status and thus more comfortable enjoying each other's society.[63]

Substance of a rather different sort stole the spotlight at another sort of company. Grand affairs—balls, assemblies, and other large parties—entranced young people with the allure of gaiety, flirting, and high fashion. At these affairs the gentry displayed its affection for what was called "dissipation": guilty pleasures whose only attraction was pleasure. Unlike salons, where constructive activities such as musical performances and conversation abounded, these affairs promised few benefits. To the contrary they were widely assumed to be detrimental, diverting youth from more responsible pursuits such as learning and piety. Anticipating her granddaughters' debuts in Charleston society in 1811, Alice Izard "long[ed] to know how they like the gay world, & whether it has many allurements for them." She conceded "the joys of dissipation," but hoped that its "va[nity] & foolish[ness]" would foster an appreciation for more sober pastimes.[64]

Given their mother's history, this was unlikely. Reading the letters her mother had written when she was pregnant with her first child, Harriet Manigault thrilled to learn that "only thing that seems to distress the young lady, is, that she is prevented from *dancing.*" Her friend Josephine du Pont was of the same mind; when home renovations prevented her from entertaining for several months in 1800, she pleaded to her friend, "I really need a little dissipation." Margaret did lose some enthusiasm for the beau monde as she reeled from the deaths of her husband and several children early in the century. But she always understood its attractions for young people and would not withhold from them the joys she had experienced as a belle.[65]

The Manigaults attended more parties than they threw, since their Spruce Street quarters were too small for large companies and Clifton was often rented out. Reciprocity demanded that they entertain, however. The salons accommodated to some extent, but the family did cram sizeable companies into their rooms once or twice each winter. Though these affairs were, naturally, more Gallic in tone than those of their American friends, they all shared common qualities. They were often quite large; Harriet Manigault noted that space limitations at her mother's house made it impossible to invite all of their acquaintances to a "large tea party" in 1813, so they settled for "between fifty & sixty" guests. Large affairs, such as the Philadelphia Assembly, might have well over one hundred invitees. Dancing was nearly universal, except at affairs such as dinner parties, and even then it might occur; the guests at Harriet's "Tea Fight" grumbled because they "expected that it would be a dance." Musicians led guests in multiple, elaborate group dances such as cotillions. On at least two occasions during the winter of 1813–14, Harriet joined seven partners in ten such dances. Passing out was common, so hosts kept on hand "all sorts of restoratives for the ladies who should faint." Women attended with a "protector"—an older brother or an uncle—as a matter of course, but as a practical matter they were largely unsupervised. The accommodations were likely to be posh as well. Women and men alike enjoyed liquors and wines. At one "charming" and "sociable" affair in 1814, Margaret noted, the guests "had a great deal of wine." Hosts offered their guests a wide variety of food and drink, and even a ball might feature a number of courses served buffet-style. A *"great tea party"* in 1813 offered guests "cakes, sweet meats, [and] Ices." Hostesses could, if they wish, make the repast a central feature of their affairs, seeking to make an impression with their exotic taste. The Hamiltons took this tack at an 1821 party at The Woodlands that "abounded in every rarity & delicacy."[66]

Underneath the veneer of tea, wines, dancing, and fainting lurked a powerful undercurrent of competition. The gentry competed indirectly with their social inferiors, both those who ridiculed them for their pretensions and those who themselves aspired to genteel status. The spectacle of four wheelers, lavish gowns, and grand houses was designed to intimidate this rabble, though its effectiveness seems to have been diminishing rapidly in the postrevolutionary decades. But the gentry were also competing among each other. Measuring one's rank was, to say the least, an inexact science, a subjective balance of wealth, family connections, political and social clout, and personal qualities. The fortunes of the Biddle and Dallas families were rising during this period, and some of the means they sought to enhance their position are evident in the observations of the Manigault family. At one 1816 party Margaret observed the company dividing between the Biddles, their "appendages," and another circle centered around the Onises. At a party three years earlier, Miss Dallas scored points by "every moment . . . say[ing] something very droll & set[ting] us all laughing really heartily."[67]

Around Philadelphia the competitive aspect of these affairs was accentuated by the presence of the European diplomatic corps. Aristocrats represented to the American gentry the epitome of refinement. The Manigaults and their like strained to show that Americans could be every bit as refined as titled Europeans. In commenting on one of their grand affairs, given late in 1813, the Manigaults revealed a sensitivity to the various levels of display that lay at the heart of early national sociability. James Cuthbert and the Wiggins family passed the test. The former, who arrived "in full dress . . . . looked very *genteel.*" Mrs. Wiggins was "splendidly arrayed in white satin, beautiful lace, & handsome pearls." Jane Biddle, however, made a lackluster impression, sporting a "common velvet dress." Her failure—which in such a prominent lady was national, not merely individual—was magnified by the scintillating appearance of the European guests. Madame Kantzow's "resplendent charms were all displayed, & ornamented with a handsome neat lace of large amethysts & a cross of Malta," while "Mme de Onis was cloathed in white satin—a rich lace robe over it—Beautiful Print lace rounded her *profuse* bosom & a little lace band modestly tied round her throat." Biddle's appearance embarrassed the Manigaults. Sumptuous affairs demanded that the gentry summon all of their resources to display their qualifications for leadership at home and respect in the Atlantic world. If they failed, why should ordinary people defer to them? Why should Europeans think of Americans as other than "a little, mean, despicable people?"[68]

American gentlefolk agonized over their precise relationship to European-style gentility. But republican aristocrats such as the Manigaults displayed little ambivalence toward grand entertainments themselves. The adults feared that immersion in the beau monde might damage their children morally, but they never considered barring them from it. "It seems to me that where pleasure, or dissipation is the object," Alice Izard reflected during a snowy March in 1811, "no weather keeps Ladies at home." In fact, the family reveled in these entertainments, the extravagance of which critics branded as un-American. Indeed, the Manigaults appropriated that word, inverting its negative connotations into a celebration of aristocratic decadence. "We are going to be very dissipated soon," announced two of the Manigault girls in early 1813, as they anticipated a season of dances and gay parties. The emerging middle class lauded restraint and self-denial, but grand affairs exalted displays of luxury and the enjoyment of sensual pleasures.[69]

The Carolina Row established a strong foundation for relations between southern planters and their peers in Philadelphia. It did so in a direct way—the Fisher family's close relations with the Middletons and other South Carolina families emerged from their participation in these salons—but also indirectly. The presence of a prominent colony of South Carolinians gave Philadelphia high society the southern cast that so impressed travelers in the decades preceding the Civil War. The Carolina Row, however, did not fare as well. Margaret Manigault died in Philadelphia in 1824, survived by only three of her children. Alice Izard lived on for eight years until she too died in Philadelphia. Her children and grandchildren returned to planting in South Carolina. Though members of the far-flung Izard and Manigault clans returned often to Philadelphia and other points north, they did so as strangers, not as sojourners returning home.

As reactionaries who unabashedly admired aristocracy, the residents of the Carolina Row hardly inspire sympathy. But they did fight for what seemed to them to be the right, which is more than their progeny could claim. The Carolina Row might have stood for later generations of gentlefolk as a model for how to maintain aristocratic standards in a democratizing society, if they had been willing to use it. But whereas Margaret Manigault and Alice Izard had set up their salon as a rebuke to the liberalizing direction of American society, their descendants used their parlors as a refuge from those same trends, which had quickened into a flood by the antebellum period. The Manigaults resisted the efforts of the rising middle class to redefine gentility, but their followers—epitomized by the Fisher and Middleton families—no longer contested the culture wars.

2

The Fisher Family Circle

THE ENGLISH SOCIAL CRITIC Harriet Martineau assessed in *Society in America* (1837) how well the young republic lived up to its idealization of equality. She found Americans obsessed with money making and unconcerned with inconvenient facts that contradicted their egalitarian ideals. She also came across women and men who, like the circle around the Carolina Row, fancied themselves aristocrats of a sort. These gentlepeople, who might have seemed threatening to republicanism, instead inspired disdain. "The republic suffers no further than by having within it a small class acting upon anti-republican morals," she wrote. Reactionaries could exercise little influence when practically every state practiced universal suffrage for white male citizens. Besides, America's mass culture, which had so impressed Alexis de Tocqueville just a few years before Martineau's travels, marginalized those who claimed special talents by virtue of their birth and social position. "The chief effect of the aristocratic spirit in a democracy is to make those who are possessed by it exclusives in a double sense; in being excluded yet more than in excluding," she observed. If such people opted to become America's "perverse children, instead of its wise and useful friends," they hurt only themselves.[1]

Had they lived to read Martineau's travelogue, Margaret Manigault and her mother would no doubt have been puzzled by the Englishwoman's distinction between "useful" and "aristocratic" behavior. As the organization of their salon reveals, they assumed that the practice of gentility was itself a public service. Reactionaries such as the Carolinians had provoked the ire of the Jeffersonian press, but in the decades preceding the Civil War they confronted a new, more insidious challenge to their public roles. Although their aristocratic pretensions still invoked ridicule, their cultural authority came under question as middle-class women and men appropriated the code of gentility for themselves. The middle class did not adopt Philip Stanhope, Lord Chesterfield's standards wholesale. Instead, they modified the aristocratic codes, replacing elitist principles with their own values, such as piety, self-control, and domesticity. Thus, the middle-class adoption of gentility, which might have invested the gentry with new authority, actually undermined it. Privileged folk who remained wedded to the older code became anachronisms—in Martineau's words, "perverse children" who forfeited their legitimacy by their loyalty to an outdated, un-American code of behavior.[2]

Cousins Sidney George Fisher (1809–1871) and Joshua Francis Fisher (1807–1873) personified the phenomenon Martineau documented. They were the very models of upper-class angst. Philadelphia gentlemen with strong family and social ties to leading southern families, the Fishers were committed to precisely the kind of social leadership to which the Carolina Row aspired. In fact, as a young man J. Francis Fisher (who was known familiarly by his last name) was an intimate member of the Carolina circle via his uncle and aunt George and Sophia (Fisher) Harrison. Fisher and his cousin sought to live up to their elitist principles in a variety of ways. As gentlemen ought, they tried to influence public opinion by writing anonymous articles and pamphlets on the major issues of the day. When some merchants urged him to expand his article "Kansas and the Constitution" into a signed pamphlet, Sidney "positively refused." He explained, "I hate notoriety of this kind."[3] Both pursued the lifestyle of country squires. Sidney ran (badly) his Mount Harmon farm in Cecil County, Maryland. Joshua erected Alverthorpe, an Italianate villa, in Montgomery County, Pennsylvania, in 1852. They also lent their time to deserving social reforms, Sidney with the Athenian Institute, a literary society, and Joshua with the Pennsylvania Institution for the Instruction of the Blind. Yet the Fishers' departures from the model set by the Carolina Row are far more striking than the similarities. They recognized the same evils the Carolinians discerned, but they were unable, and even unwilling, to do anything about them. Where the Izards and Manigaults had conceived of their parlors as a bulwark against democratization, the Fishers' circle saw theirs as a retreat from these same forces. Lacking the confidence or gumption to fight, they conceded the battlefield.

At the same time the Fishers had to contend with a force that troubled the Carolinians hardly at all: sectional antagonism. The social and family lives of the Fisher cousins underscore how allegiance to a common, increasingly anachronistic standard of refined behavior forged the northern and southern gentry into a unified leisure class well into the nineteenth century. But it also demonstrates how conflicts over the extension of slavery crossed the political threshold to transform American society and culture. While J. Francis Fisher identified ever more closely with the southern planter class, in the 1850s Sidney Fisher grew more hostile toward the South and slavery. Vices that had seemed national in 1830 appeared to be distinctively southern in the light of the sectional crises of the final decade of the antebellum period. Though Sidney struggled mightily to compartmentalize politics and social life, his disdain for the South's political course eventually tainted his opinion of southern men and women. By contrast J. Francis Fisher's allegiance to the South was viscerally personal; to repudiate it would have meant severing family ties as well as forsaking the reactionary convictions that allowed him to conceive of himself as an American aristocrat. The Fishers' agonies underscore the dilemmas that the intersection of sectionalism, nationalism, and gentility in the mid-nineteenth century posed for privileged Americans.

". . . our circle in Philadelphia"

The Fishers were the first generation of their families to establish warm bonds of kinship and friendship with southern families. These linkages were the products of the same nationalistic sentiments that brought the Manigaults to Philadelphia and rendered them comfortable there. Their reactionary sentiments, however, had deep roots. Both Sidney and Joshua were the descendants of James Fisher, a Quaker who had immigrated to Pennsylvania with William Penn. Sidney and Joshua's ancestors were among the city's leading dry-goods merchants. The family remained socially prominent and economically secure despite their loyalism during the American Revolution. Sidney's parents, James L. and Anne Elizabeth (George) Fisher, raised their children in comfortable gentility. Sidney was bedeviled throughout his life with the awareness that he lacked the wealth to match his social position. After a family dinner late in 1843, he took comfort in the knowledge "that tho' my family is neither so rich nor so influential as it once was, I come on both sides from good blood." Though he lacked the riches that the other members of his circle possessed, his social position was never in doubt. He kept an active social calendar in the city and in national centers of gentility such as Newport. Thus Sidney felt justified "to consider that 'I am a gentleman,' tho' a poor one."[4]

His cousin was more fortunate. He enjoyed the same prominence in society but possessed riches that were Sidney's envy. His father died just a few months after his wedding to Betsey Francis, leaving her saddled with debts. Because Joshua's mother came from Philadelphia's non-Quaker gentry, his father's relations shunned her and resisted paying to him the inheritance left by his grandfather. Young Joshua and his mother were taken in by her childless sister, Sophia, and her wealthy merchant husband. George Harrison had been a protégé of Robert Morris, the Revolutionary financier, and had made a fortune as a wine merchant. He raised Fisher as his own son. He taught him to manage his finances, paid his way through Harvard, sent him on a Grand Tour following graduation, and devoted special attention to molding him into a gentleman. On Fisher's marriage to Eliza Middleton in 1839, the Harrisons guaranteed him an income of nearly $8,600. Although the young man anticipated devoting himself to pursuing the life of a gentleman of leisure and "indulging [Elizabeth] in every reasonable wish and taste," in fact Fisher proved to be as skillful an investor as his uncle. He increased his fortune considerably by dint of his own efforts over the next two decades. Unlike his cousin, Joshua enjoyed opulence commensurate with his social station.[5]

Both Fishers were inveterate snobs. Raised within the city's Federalist establishment, they developed a profound distrust of the capacities of ordinary people. Only men who possessed family eminence, advanced education, economic independence, and social standing were fit objects of respect. As Sidney contended, "I always feel socially superior to a man who is not a gentleman by *birth,* and I never yet saw one who had *risen* to a higher position, whose mind and character as well as his manners did not show the taint of his origin." Joshua shared these elitist sentiments. At Harvard his friends "kept company with each other in aristocratic exclusiveness, into which only a few exceptional Yankees of families of high distinction were admitted." He mocked the notion that "poor people would be found actuated by the same high notions" as "the richer classes." Other figures in the Fishers' orbit, such as Roberts Vaux, Charles Ingersoll, and the Butlers, made their peace with democratization. But the cousins never reconciled themselves to these changes and viewed those who had as apostates. The best Sidney could say about Pierce Butler, "a Van Buren man," was that "the more men of property they have the more conservative [Democrats] will be." Yet the Fishers never seem to have accepted the marginal status to which their reactionary principles fated them. Writing to Joshua in Paris in 1832, Sidney observed that "to succeed in commanding attention & engaging interest, in a distinguished & brilliant circle is to gratify no insignificant ambition." Yet the methods necessary to gratify ambition in Jacksonian America were exactly those that the Fishers judged to be beneath their dignity. Thus, their conservative attitudes bore not only the fruits of irrelevance, but of bitterness.[6]

Though Joshua had closer family ties to the South, Sidney's were not inconsiderable. After a happy bachelorhood, he married Elizabeth (Bet) Ingersoll in 1851. Bet had grown up surrounded by southerners and by southern sympathizers. Her father, Charles Jared Ingersoll, was a Jeffersonian congressman whose widely read *Inchiquin: The Jesuit's Letters on American Literature and Politics* (1810) made the case for cultural nationalism. Throughout the sectional crisis of the 1850s he supported the South's efforts to extend slavery into the territories. In 1861 he endorsed secession, not only for the South but for Pennsylvania too. Antislavery meetings and crowd actions against them were common events in antebellum Philadelphia. It is certain that their occurrence were prime topics of conversation in the Ingersoll household. Bet's friends also had southern roots. She was close with her brother's wife, Susan Brown, the daughter of Dr. Samuel Brown, a prominent southern physician. Charlotte Manigault Wilcocks, Margaret Manigault's granddaughter, was among Susan's closest friends. By virtue of her friendship with Charlotte, Bet entered the society of the Izard and Butler families. Before their marriage Sidney and Bet Ingersoll moved in the same circles, but their union drove him even closer to prosouthern elements in Philadelphia.[7]

J. Francis Fisher's ties to the South were far stronger than his cousin's because he married into one of the region's most prominent planting families. On March 29, 1839, he and Elizabeth Middleton exchanged vows at Middleton Place, her family's estate outside Charleston. Elizabeth was the youngest daughter of Henry and Mary Hering Middleton. Her mother hailed from a respectable English-gentry family, and her father was the oldest son of Arthur Middleton, a signer of the Declaration of Independence. By the Revolution the family was among the wealthiest in South Carolina, and the cotton boom of the early nineteenth century only expanded their fortunes. Eliza and Joshua were a good match. Sidney, who had an eye for such things, observed that she "has no beauty and has red hair, but her countenance has a pleasing expression and her figure is good."[8] Fisher was not known for his looks either, but a portrait completed about 1835 hints at his active intellect, self-possession, and sense of style. Both he and Eliza shared a taste for music, late-night socializing, and dance—not surprising, since it seems the two had met at the fashionable resort city of Newport. She complemented his charitable activities by her patronage of visiting musicians.

The families were in some ways better matched than the principals. Fisher quickly became a favorite of Eliza's parents and brothers, particularly her brother Williams (1809–1883). Through Joshua the Middletons came to meet

Sidney, whose eccentric personality and reactionary politics they found irresistible.[9] Philadelphia became a kind of second home for the Middletons after their alliance with the Fishers. Eliza's brothers visited the city during the social season. They vacationed in Newport during the summertime with other privileged families, particularly after 1842, when Eliza's relation Nathaniel Middleton married Anna De Wolf of Bristol, Rhode Island, whose house became the nexus of their Philadelphia-Charleston-Newport social circle.[10]

These family ties firmly oriented the Fishers' social circle toward the South. It did not create that orientation, however. The foundation of the close relations between southerners and Philadelphia families such as the Fishers are to be found in preexisting social webs and class attitudes. People of privilege established connections across state and regional lines in the postrevolutionary period. In part this was a product of nationalist sentiment, as Americans responded to a widely felt imperative to create nationality by fostering a sense of national community.[11] This impulse, together with other factors, fostered a boom in domestic tourism in the early national period. Travel promoted interpersonal connections as like-minded people sought each other out in public spaces and social occasions.[12] Many members of the Fishers' social circle—such as the Biddle, Butler, Rush, and Middleton families—had been intimates of the Carolina Row. The Fishers' set was an outgrowth and elaboration of that earlier collectivity.

The many differences between southeastern Pennsylvania and the South Carolina lowcountry notwithstanding, the Fishers and their southern friends and relations shared much in common. Wealthy and conservative families from Newport to Savannah subscribed to a common code of refined conduct. Most important, they enjoyed wealth and privileges that few Americans of their generation could imagine. On his death in 1845, George Harrison left an estate valued at $470,000. The Harrisons and their circle constituted a leisure class, aspiring to adopt a style of life modeled after the eighteenth-century English gentry. They did not glorify work, thrift, self-control, or self-sacrifice. Nor were they devoutly religious. In fact, they had little but contempt for demonstrative expressions of piety. Mary Hering Middleton was a diligent Episcopalian and raised her daughters accordingly. Few of her sons were more than indifferent to religion. Angry over how his father's relations snubbed his widowed mother, Joshua Fisher had no use for the Friends, though he too attended Episcopal services. To his mind religion's most useful function was social control. While in France he even found himself arguing with Marie Joseph du Motier, Marquis de Lafayette, over the subject. Fisher insisted that "religion was necessary for the base people, to keep them in good order & preserve their morals." In this respect, as in many others, these families distinguished themselves from ordinary people both in the North and South. Moreover, their latitudinarianism set them apart from an increasing number of planters as well.[13]

As Fisher's dispute with the Revolutionary hero attests, some gentlepeople found religious devotion distasteful precisely because they associated it with low social status. The poor needed piety to deflect them from vice; refined people possessed qualities such as "honour, generosity, & patriotism" that allowed them to enjoy pleasures without descending into viciousness. The Fishers' southern circle engaged in social activities that respectable people would have condemned as debauchery. As Lydia Maria Child, an advice writer for families "of moderate fortunes" wrote, "the prevailing evil of the present day is extravagance." Card playing, waltzing, alcoholic consumption, and entertainments that lingered into the wee hours were de rigueur for the gentry. In 1822 Mary H. Middleton told Septima Rutledge that families—hers included—went "night after night to balls which lasted till 2 or 3 o'clock in the morning." During the winter of 1838 Sidney Fisher attended a party at George Cadwalader's house, "whose rooms are very sumptuously furnished & are decidedly the handsomest in town. The walls are beautifully painted in fresco . . . the chairs white & gold, and there was a profusion of splendid candelabra, vases, etc., so that the effect when lighted was very rich and beautiful." Quasi-aristocrats such as the Middletons and Fishers assumed that self-indulgence on such a scale was their birthright. Refined surroundings enhanced the character of mundane activities such as card playing and dancing, the gentry believed. They also assumed that only privileged people possessed the character to indulge in these behaviors without becoming corrupted. Environment could only partially efface such capacities, even over time; they were, almost in a racial sense, innate.[14]

Education reinforced these similarities of wealth and lifestyle. The upper ranks of society in both the North and the South continued to patronize private academies and elite colleges and universities, where they learned habits of mind that crossed sectional boundaries and forged friendships with young people from across the Union. Eliza's brother Arthur graduated from Harvard in 1814; Fisher earned a bachelor's degree there eleven years later. Harry Middleton graduated from West Point in 1815, and John Izard Middleton took his degree from Princeton in 1819. Social bonds formed outside the classrooms, particularly in the private clubs that proliferated on antebellum campuses, became intersectional friendships after graduation. At Harvard, Joshua

Fisher and his friends organized the Porcellian Society, which restricted membership to students from elite families, preferably from outside New England. Among its members were his lifelong friends Reverend Paul Trapier of South Carolina (who officiated at Fisher's wedding) and Bostonian Allyne Otis. Southern parents were mindful that their sons might encounter sentiments hostile to the peculiar institution in the North, but these fears seemed minor compared with the benefits to be gained by the superior quality of northern colleges and the social connections to be made there. Besides, most urbane planters were staunch nationalists well into the 1850s.[15]

Elite women's education also reinforced intersectional bonds, as Eliza Middleton's case illustrates. Her schooling did not prepare her to engage in productive activities outside the home. Like her brothers, she studied French, drawing, dance, voice and musical instrumentation, and other "accomplishments." Like them, she received instruction in formal subjects such as literature, history, and the sciences, not for their application in the workplace, but because they were essential skills with which to shine in company. Eliza trained as a vocalist and played several musical instruments with considerable skill. In 1822, when she was six, her mother wrote to inquire about her studies. "And music!" she added after urging her to master the French, German, and Russian languages. "I wish to hear how my dear little girl is coming on in that." Following her mother's wishes, Eliza spoke several languages fluently and could read still more. She was well-read, sociable, and a gifted conversationalist—skill at conversation being a particularly prized accomplishment. Joshua's cousin found her to be "clever, cultivated, accomplished, and agreeable." In fact, it was her sophistication that won Fisher's affection. His first view of Middleton Place, which bore "more of the signs of civilization than any thing I have seen in America," confirmed his early impressions of Eliza's refinement.[16]

Finally, all the Middletons and J. Francis Fisher (but not Sidney, to his chagrin) spent extensive time during their youths in Europe. Overseas travel was not mere recreation. It was also an educational endeavor—an exercise in "the study of man under various aspects," as Sidney told Joshua in 1832. "Besides, there is some excitement in it." Perhaps more than any other feature of their education, European travel helped them transcend regional differences to promote a sense of belonging to a national (and nationalistic) privileged class. Several of Eliza's brothers attended universities in Edinburgh and Paris. All of them learned from private tutors in England and in Russia while their father served as minister at St. Petersburg during the 1820s. All save Catherine, who suffered a mental breakdown and was institutionalized at Pennsylvania Hospital, later enjoyed extended travel on the Continent, spending months on

English-style Grand Tours in the major cities, where they mingled with well-born Europeans and visiting Americans.[17]

They also served their country as diplomats. Joshua assisted the American minister in Paris, William Cabell Rives, for a time, and several of the Middleton boys served under their father. Arthur, Edward, and Harry Middleton took European brides. More important, their travels intensified both their elitism and their identification with America. While on his 1838 tour, Henry Middleton complained about the "incurably underbred habits and vulgar tone" of Americans in Paris, which he attributed to the paucity of "*gentlemen* and men of education" from the States. "It is in fact that very feeling of patriotism which suggests what I say," he explained. "It is because I feel proud of America that I wish to see her well represented." At his uncle's insistence Fisher toured England, France, Switzerland, and Italy in 1830–32. Travel affected him in the same way as it had the Middletons. He faulted Judith Rives for being a "little *too* American" for her lack of "ease & tact." Overall his affection for his native land intensified. "I have seen in Paris no more elegant men or women in dress, manners, or mind that I have met in America," Fisher wrote in a typical passage. Travel deepened the patriotism of the Fishers' circle, but it also distinguished them from their country people in important ways. Only the wealthy could afford overseas leisure travel, and only the most privileged could afford the English-style Grand Tour the Middletons and Joshua Fisher enjoyed. The experience of visiting the Old World invested travelers with a special authority. Travel also afforded exposure to real aristocrats, who provided models of behavior unavailable to all but a few Americans.[18]

These similarities forged the Middletons and Fishers into an extraordinarily close alliance. The Fishers made several trips to Middleton Place and Charleston, and the Middletons were de facto Philadelphians after the 1839 wedding. Nearly a decade after the ceremony, Williams Middleton took a break from supervising rice planting on his Combahee plantation—he called himself "Inspector General of plantation matters"—to thank his brother-in-law for a letter passing on "favorable news from our circle in Philad[elphia]" and expressing anticipation "with no little pleasure to the time now not many weeks distant when I may rejoin you all." The Middletons increasingly took to addressing their letters directly to Fisher, though not because of any sense of patriarchy. (Eliza offered candid advice to her brothers as she saw fit.) Rather, they looked to Fisher as the northern nexus of the family. In 1840 Williams thanked Fisher for "our last news from Philadelphia," since he had received not a single missive from his brothers or sisters—then traveling in the

North—for weeks. "The other members of the family appear to have forgotten us," he despaired.[19]

"the exclusives"

Kinship ties made the Philadelphia Fishers southern in temperament. The social circle in which the Fishers circulated was also oriented toward the South. Like other American cities, Philadelphia's upper crust was divided into numerous overlapping cliques defined by family allegiances, standards of behavior, religion, and neighborhood.[20] The social set with which the Fishers were most closely identified was the one associated with reactionary sentiments (Eliza Fisher called their circle of families "the exclusives") and southern orientation. Among their closest associates were the Butlers, a family with strong roots in the deep South. Pierce Butler, a major in the British army during the French and Indian War, settled in the South Carolina and married Mary Middleton in 1771, soon becoming one of the lowcountry's leading rice planters. In the 1790s he decided to establish residence in Philadelphia. Butler bought a town house on the corner of Chestnut and Eighth streets as well as two rural properties outside the city. The downtown mansion and Butler Place, as one of the rural houses became known, were centers of prosouthern sociability in pre–Civil War Philadelphia.[21]

The major's grandsons, John and Pierce Butler, were close members of the Fishers' circle. Their mother, Sarah Butler, married James Mease, a Philadelphia doctor from a well-connected family. Their material grandfather required that they take his surname if they wished to inherit his Georgia estates. They did so, thereby entering two exclusive cohorts: they were among the wealthiest men in Philadelphia, and they ranked among the South's major slave owners. One exposé of the city's high society revealed that John Butler had "married a southern lady" and "was supposed to be worth at least" $50,000. Another gossip sheet put John's wealth at three times that sum and estimated that his aunt Elizabeth, the major's daughter, was worth $150,000. The Butlers' wealth, breeding, and sense of entitlement marked them as gentlemen of the first order. When Sidney Fisher visited Butler Place for the first time in 1839, he praised Pierce for being "very gentlemanlike in manner." The house and its grounds also bore the marks of cultivation. Butler Place was "surrounded by fine trees & the grounds are admirably kept." These high opinions did not last, but the Fishers' conviction that the Butler men were gentlemen who deserved the rank they had acquired by renouncing their patrimony for their grandfather's fortune never waned.[22]

Upper-class women bore much of the responsibility for maintaining their families' social position, and the Butler ladies were no exception. In fact, they were held in higher regard than their husbands. Fanny Kemble married Pierce Butler after a whirlwind courtship in 1834.[23] The daughter of English thespian Charles Kemble and herself an acclaimed actor, she had been touring America when she met Pierce Butler in 1832. The match was an impulsive decision that they both came to regret. Pierce came to understand that, at almost every point of comparison, he was diminished by his wife. He resented her fame as a writer and the esteem in which she was held by Philadelphia society. She hated slavery and, worse, publicized her views by the publication of her American *Journal* (1835).[24] Fanny's "noble qualities," as a Philadelphia friend called them, highlighted her husband's flaws. Besides possessing a conventional mind, Pierce had a passion for extramarital affairs and gambling, the latter causing near-constant financial embarrassments. When Butler's niece confided to Charlotte Wilcocks that he "kept a mistress," the young socialite resisted the urge to tell her friend that "her Uncle kept two." They separated in 1845, when Fanny returned to England without her daughters. They finally divorced in 1849.[25]

Though even her friends criticized Fanny for being too masculine—a common slur against intellectual, opinionated women—she remained in the good graces of her old circle. Sidney condemned her for being "the reverse of feminine in her manners & conversations," but he continued to treasure her "qualities of heart & character," which he judged to be "as excellent as those of her intellect." Eliza and Joshua Fisher rated her friendship so highly that they remained close to Fanny in spite of her antislavery convictions. One morning when they "nearly got drawn into a discussion on Abolition" Fanny kept the peace by confessing that she published her 1835 *Journal*—which revealed her antislavery beliefs long before her better-known *Journal of a Residence on a Georgian Plantation* (1863)—only because Pierce risked defaulting on a debt. Whether within the Butler fold or estranged from it, Fanny Kemble was an integral part of the Fisher family's circle.[26]

Gabriella Morris Butler, John Butler's wife, lacked Fanny Kemble's wit. To the cultivated Eliza, she was at best "goodnatured & amiable"; more often, she was "neither talkative nor interesting." However, she complemented Fanny well because she possessed qualities that her sister-in-law lacked. Her pedigree was impeccable. Margaret Manigault was her grandmother; Lewis Morris and Elizabeth Manigault Morris of New York were her parents. As a girl she summered often with her aunt, Harriet Manigault Wilcocks, at her Philadelphia home, where she met John Butler. While she did not possess wit or the gift of sparkling conversation, she was endowed with great beauty and a sense of refinement that allowed her to dominate a space by her presence alone. Even when she dressed in the garb of a peasant girl at a masquerade ball in the winter of 1846, "her appearance was as good as any in the room" to Sidney Fisher's practiced eye. Her beauty was such that it was "more conspicuous in a simple dress than in any other," although he conceded "it is also well suited to a rich costume." To Sidney it hardly mattered that "she ha[d] no mind," since she possessed "the ease & high breeding of the southern aristocracy, a manner produced by birth, early habit & wealth."[27]

Besides the Middletons, Fishers, and Butlers, there were other families with southern ties in the highest echelons of Philadelphia society. William Drayton (1776–1846) moved his family from South Carolina for Pennsylvania after the triumph of pronullification forces. As the last president of the Second Bank of the United States, he enjoyed considerable prominence in the city's Whig establishment.[28] Also within the Fishers' circle was the Wharton family, led by Thomas I. Wharton (1791–1856), the first president of the Historical Society of Pennsylvania. The Whartons were deeply prosouthern. During a trip through the South in 1853, Joseph Wharton concluded that "the position of the slaveholders is one of much delicacy and not at all likely to be improved by intemperate urging for abolition."[29] This temperamental conservatism was reinforced in 1842 by the marriage of Thomas's daughter Emily to Charles Sinkler, a naval officer from a wealthy cotton-planting family of upper St. John's Parish, some sixty miles from Charleston. When Emily prepared to make a visit to Philadelphia in 1843, she arranged with Eliza, then visiting Middleton Place with Joshua and their children, to return with her. Thomas Wharton was active in Philadelphia's conservative social circles; like Sidney and Joshua, he was a member of the exclusive Wistar Association, the social arm of the American Philosophical Society. The Sinklers were on close terms with the Middletons. The families exchanged visits when they were in Charleston.[30]

Because of the strong southern presence in Philadelphia's high society, visitors from that region were an ubiquitous presence at affairs given or attended by the Fishers' clique. Moncure Robinson, the Virginia scientist, invited Sidney Fisher to a party in 1837. "All the Virginians in town were there as a matter of course," he observed. Two years later he attended a party with Hugh Swinton Legaré, a South Carolina man of letters. Fisher praised the Carolinian's "talents, literary acquirements, [and] agreeable powers in society." He was the very model of the republican aristocrat, "one of the few eminent men left who combined talents of a high order with the culture and attainment of a gentleman." Sometimes so many southerners were in the city that they outnumbered the natives. At a wedding party in honor of Dr. Charles Carter, the Virginians—including Carter, George Tucker, Edward Coles, William Cabell Rives Jr., and William Byrd Page, the last a medical student at the University of Pennsylvania—eclipsed the Philadelphians. Being surrounded by southerners presented no obstacle to enjoying the evening. "Got along very well," Fisher noted contentedly before retiring that evening.[31]

Befitting their social leadership, southern women also occupied a rarified place in the Fishers' circles. At a small party in 1841, Sidney listened with rapt attention to Sallie Coles Stevenson, wife of a former minister to the Court of St. James. He particularly liked her stories about "the awkward behavior and mortifying exhibitions of some Americans on being presented at court," conduct that reinforced his notions of American vulgarity. He rated Stevenson "a great talker, with all the Virginia fluency."[32] In 1847 Sidney spent an evening with Margaret Telfair and her new husband, William Hodgson, at a party given by Sarah Marshall Atherton, herself a southern lady. The affair was in honor of the visiting Georgians, whom Humphrey Atherton had befriended the previous summer. Eliza Fisher enjoyed the company of Maria Vidal Davis, the wife of Natchez planter Samuel Davis, who shared Eliza's taste for vocal music. Phoebe Rush's weekly salons and her yearly ball, a social highlight of the season, included the women of the Izard, Middleton, Drayton, Telfair, and Carter families when they were in the city. Via its control of the highest echelons of Philadelphia society, the circle to which the Fisher's belonged gave the city a southern orientation.[33]

". . . a duel would be quite necessary"

Philadelphians did not open their doors to visitors from the slave-owning regions unless these women and men possessed the necessary connections. Because of this exclusivity, relations between planters and their northern peers were extraordinary close during the 1830s and 1840s. But these personal connections began to unravel as the sectional conflict over the extension of slavery into the western territories intensified after the Mexican War (1846–48). Women and men had long clashed over public policy and candidates for office, of course. As the issues became more divisive and bitter, however, these differences began to erode friendships. Sectional peculiarities, such as accents and small points of manners, came to be seen less as quirks than as moral deficiencies. Despite these conflicts, the Fishers' circle remained staunchly southern in temperament during the antebellum period. A sense that the sectional conflict was a product of excessive democratization tempered antagonisms. If the struggle between the North and South exacerbated certain fault lines within the Fishers' society, it strengthened others.

Some Philadelphians recoiled from what seemed to be a greater tolerance for interpersonal violence among their southern peers. Dueling, cockfighting, crowd actions, and other forms of violent behavior steadily lost their legitimacy in polite circles in the North during the first half of the nineteenth century.[34] However, these behaviors also came under attack by middle-class activists in the South.[35] Religious critics in both sections agreed that they violated Christian morals. Exchanging gunfire and street fighting were hardly consistent with the canons of good taste either. William Grayson of South Carolina charged that dueling was "the product of a barbarous age and flourishes in proportion as the manners of the people are coarse and brutal." Nevertheless dueling persisted in the slave states. Worldly planters saw it as a genteel way to settle disputes between gentlemen. In the Deep South underdeveloped social institutions gave license to an ethic of unrestrained masculinity.[36] Honorific violence of any sort invoked the censure of the vast majority of northerners. In 1849 a young friend of Rebecca Gratz received a challenge from a hot-blooded German immigrant, who claimed that he had maligned his native land by ridiculing waltzing. The charge was groundless, but the Philadelphian avoided fighting a duel only by thrashing the hapless young man on the street. Gratz found no humor at all in the Clay-Turner duel of 1849, in which both men were killed. The affair exposed the consequences of associating gentility with dueling. A devotion to "honor, false honor against every principle of right" misled the men to "mingle wrong & right together to do a wicked thing to prove that [they] are brave."[37]

Dueling was more widely denounced in popular culture. In 1826 the *Philadelphia Album,* a women's magazine, waged a frontal assault on the widespread association between dueling and refinement, which threatened to make the practice popular with social-climbing young men. "Harriet" portrayed the duel not as the embodiment of cultivation but as its antithesis. The fatal ball, the poet warned, slew "the graceful, beautiful, and brave / He fell for honour's empty name." Later that year, a "Lady of Charleston," whose roots may have given her more credibility on the matter, condemned dueling in similarly sentimental terms. "In this enlightened age," she wrote, "we no longer consider success (the soul shudders at the term) as the test of innocence." No man should "sit down tamely under his wrongs"; rather, there ought to be established a "Court of Honor" that would adjudicate "those injuries and insults which are now submitted to the Court of Death." Opponents of dueling, North and South, confronted an evil they characterized in class, not sectional, terms.[38]

The reactionary northern gentry, including the Fishers' circle, were themselves prone to the same fits of intolerance and "refined" aggression they claimed to abhor in their southern friends. Sidney Fisher condemned the South for its "ferocious & vicious manners, a low & degraded standard of morals and opinion." Yet he was more than willing to countenance communal violence, particularly when it was directed against abolitionists. Philadelphia socialites excommunicated radical critics of slavery from their world. Zeal of any kind made for poor conversation, and critics of the peculiar institution had no place at all in a community so rife with slaveholding connections. In the spring of 1838 a mob consisting of "well dressed men" burnt down Pennsylvania Hall during a biracial meeting of antislavery women. Sidney observed that the police declined to disperse the crowd, which prevented volunteer fire clubs from fighting the fire. He deplored such disrespect for "the supremacy of the laws," though he went out of his way to excuse the arsonists. "To be sure there was great provocation," Fisher mused, for "the fanatic orators openly recommended dissolution of the Union, abused Washington, &c." But abolitionists violated more than nationalist taboos—they also crossed sacrosanct racial lines. "Black & white men & women sat promiscuously together & walked about arm & arm," he marveled. Though he could not endorse the behavior of the crowd, which had set fire to the new building while the conventioneers were still inside and tried to prevent their escape, in this case he blamed the victims for bringing such suffering onto themselves. "Such are the excesses of enthusiasm," he concluded laconically.[39]

Philadelphia gentlefolk even endorsed dueling in certain circumstances. They possessed a prickly sensitivity about their reputations. In addition, they identified antidueling sentiment with middle-class social activism, which enhanced the practice in their estimation. In 1830 William H. Keating, a Philadelphia mineralogist, complained to his friend Joel Roberts Poinsett about "much jarring in the elements of society in Philadelphia," manifested by the vigorous activities of "tract societies, temperance societies, antisabbath mails, antimasonics, &c." These social reform efforts, which Keating interpreted as unwonted intrusions of bourgeois values into the public culture, had been "given plausibility" by two episodes of genteel violence. The first was "a very fatal duel . . . which has raised great feeling in the community." The second was a public beating administered by a young man on another who "had published a pamphlet derogatory to the character of his sister." Together these incidents showed "which way the wind blows"—in favor of "religious" and middle-class interests, against the privileged orders. But even popular opinion regarding dueling was mixed. One exposé of upper-class society reported the delicious rumor that Benjamin Smith Barton, worth "$100,000," had "once killed a gentleman in a duel."[40]

Honorific violence was a complex affair, with class and sectional overtones that defy an easy explanation. A scandalous episode in the Fishers' circle underscores some of these tensions. In 1849 Edward Middleton separated from his wife of four years, Edwardina de Normann, whom he had met in Naples while serving as a lieutenant in the U.S. Navy. His family had uniformly opposed the match. Edda claimed noble lineage, but the Middletons (with good cause) doubted her credibility. They separated in 1849, when she and Edward were back in the United States. The cause was Edda's scandalous sexual indiscretions, one of which involved an affair with Harry McCall, the Philadelphia gentleman who had married Charlotte Manigault Wilcocks. Sidney Fisher heard from Thomas B. Huger, a naval officer, that Edda had "from the time of their marriage flirted desperately & most imprudently in all directions." The ferocity of the rumors compelled the Middletons to send Edda out of Philadelphia, but the scandal could not be held back. Eliza demanded her brother divorce "the vile little hussy." At the very least, she insisted, he should spare the family's reputation by keeping "his strumpet wife" away from the city. Her brother rejected these appeals. He continued to seek reconciliation with his wife even while she consorted with men of "doubtful reputation" and, as Sidney Fisher learned, took "one lover after another, was openly the mistress of the keeper of a fashionable hell in New York & passed also thro other various hands."[41]

Such unmanly behavior deeply shamed both the northern and southern branches of the family. But it was the Fishers who interpreted the affair in the language of honor and who demanded a resolution in the terms of the *code duello.* The Middletons, by contrast, urged Christian restraint. Edward sheepishly sought reconciliation with Edda in 1850, provoking Joshua to question his manhood. He ought to "assume the romantic name of de Normann," Fisher wrote to Williams Middleton. In other words, Edward should formalize his unmanning by taking his wife's name. Eliza berated her elder brother for "dishonouring us all." She and Fisher were "prepared to defend their own honour" by exposing Edda's vices if he would not defend himself. When Edda's affair with Harry McCall came to light, the Fishers assumed as a matter of course that the cuckold would demand satisfaction. Disdainfully Sidney noted that in the Southwest such a scandal would give the violated man license to "shoot or stab his injurer wherever he should find him without notice." In civilized areas such a course of action was properly "regarded as barbarous." Nevertheless, he observed, "except perhaps in Boston, a duel would be quite necessary."[42]

In addition, most of the family's social circle expected Edward to seek "revenge" against McCall. They doubted whether he could "maintain his position in the Navy or among gentlemen" unless he confronted the rake. In fact, "many severe things" were said about Edward "here [Philadelphia] & in Carolina" when it came out that he had "conscientious scruples on the subject." The Fishers were aghast, since Edward's shame tainted them by extension. What was worse, his brothers ratified his conduct. Dueling, Arthur Middleton explained, made "every man his own avenger in the bloodiest sense" and was "destructive of the very foundation of Christianity." Buttressed by the support of the southern branch of the family, Edward refused to budge. His moral convictions, the support of his brothers, and the martial spirit articulated by the Fishers cautions against regarding honor as a strictly sectional ethic. Dueling and other honorific practices were aristocratic—not strictly southern—behaviors.[43]

"Only think of my dissipation!"

Besides seeming hot-headed and prone to violence, southerners also appeared to lack refinement. As in the case of violence, however, these attitudes were complex. Northern attitudes toward the South blended repugnance with admiration. Southerners seemed uncultured, profligate, and coarse. Yet they also seemed more comfortable in their own skins. The middle-class assault on aristocratic refinement eroded the confidence of the northern gentry far more than the southern. Privileged Philadelphians could not help but be influenced by criticism of the ideal of gentility. "To be a gentleman," wrote Benjamin Rush, "takes away all disgrace in swearing, getting drunk, running in debt, getting bastards, etc." Such remarks were part of the common stock of the middle-class critique of gentility, and, as Rush's comments illustrate, they were to a significant extent absorbed by the northern gentry. But northerners were also compelled to admire southern planters for their stricter adherence to older, English-inspired forms of refined behavior that stressed the outward display of wealth and power. These tensions sharpened as political events intensified the importance of sectional differences. The resulting strains afflicted Philadelphia high society for much of the antebellum period.[44]

Antebellum Philadelphians shared the widespread northern belief—epitomized by the character of Marie St. Clare in Harriet Beecher Stowe's *Uncle*

Tom's Cabin (1852)—that southern women lacked the manners and morals of northerners. When he announced his plans to marry Eliza Middleton, Joshua Fisher assured his uncle that she had "none of the habits of a southern woman," which he identified as "indolent helplessness and languid carelessness." Sidney Fisher used the same language in judging Charlotte Taylor Robinson, the wife of a Virginia engineer. She was a "small, thin, languid-looking woman," he declared after attending an affair at her house. Sidney was in some ways charmed by the Mason sisters (also Virginians) when he met them at a party in early 1838. One of the young ladies boasted "a good figure, regular fea-tures, fine complexion, pleasant smile, & intelligent expression." Nevertheless, she was "careless in dress & person, affected, familiar, badly educated, & without the appearance and manner of a high-bred lady." Philadelphians, like northerners generally, laid the blame for these southern faults directly at the feet of slavery. Joshua bragged that his bride-to-be possessed none of the vices "characteristic of women bred up among slaves." Sidney charged that the Masons' faults were "thoroughly Virginian," endemic to all women "bred at the South."[45]

Southern men fared no better, as Philadelphians' evaluation of John and Pierce Butler testifies. Major Butler's heirs lacked the manners and morals that should complement a great fortune. In this respect at least they followed their illustrious grandfather's lead. Joshua Fisher recalled the old planter as "beyond all men violent, contrary, and tyrannical." There was scarcely any "elegance or appearance of taste in any thing about him." His grandsons seemed to grow more like him as they aged. No one characterized Pierce's conduct as a husband as anything but debauched. And while Sidney complemented his "highly honorable" efforts to prevent the sale of his Georgia slaves in 1858, he also blamed Butler's "sheer infatuation" for gambling and profligate spending for the debacle.[46] Philadelphians had a similarly dim assessment of his brother John. When word reached the city of his death while serving in the Mexican War, Joshua Fisher noted coldly, "since I was 14 I have known little of him—nor desired to." Sidney, who knew him better, said John Butler was "a hard, selfish, profligate fellow, totally without education or intellect." In these characterizations of the Butlers, the Fishers articulated the northern critique of southern manners. Southern gentility, marred by the influence of slavery, was crude, debauched, money grubbing, and patriarchal compared to its northern variant.[47]

Other southern-oriented members of Philadelphia's high society fared no better in the Fishers' estimation. Joshua Fisher recalled that George Izard, the younger brother of Margaret Manigault, possessed a "haughty demeanor" instead of the more modest pride appropriate to a gentleman. Izard, from an established family, should have known better. Samuel and Maria Davis were a different case. Relocating to Philadelphia from Natchez after securing their cotton fortune, the Davises threw a small party in 1838 to display their refinement. It was at best a mixed success. "They spend their money like all people in half-civilized countries," huffed Sidney Fisher. Their arrangement of their tasteless furniture inhibited sociability. They served food and drink in the most expensive china and glass. Designed to impress their guests, this display instead "indicates a truly parvenue spirit, which mistakes glitter and gaudiness for taste and elegance."[48] Haughty, garish, parochial, anti-intellectual—these were the key charges in the North's indictment against southern gentility. As the friends and allies of the southern planter class, Philadelphians were, willingly or not, witnesses for the prosecution.

Philadelphians' true sentiments were more complicated. Misgivings aside, they found much to admire in planter-class women. When Sidney Fisher vacationed at Newport in 1841, he ran into the Masons again. Three years had apparently affected a startling change in the Virginia sisters. Betty Mason impressed Fisher with her "regular, classic" beauty and her "graceful, easy, and ladylike" manners. Her younger sister, Matilda, also stood out among the fashionable throng as a "rosy, laughing, beautiful Hebelike girl with an exquisite figure."[49] Gentlepeople could apply these standards brutally, as when Charlotte Wilcocks ridiculed a Washington girl for having such "bad feet" that she "must have been born a scullion." Because Philadelphians were relatively untainted by antisouthern prejudice and because of such unabashed elitism, they and their planter companions looked remarkably similar to outsiders. A Boston journal charged that both Charleston and Philadelphia women were "adepts at dress and other frivolities" concerned only with "husbands who can give them fine houses, fine furniture, fine frocks, and fine bonnets." Reactionaries might have objected to that bill of particulars. But they would have agreed that an elitist sensibility bound gentlepeople together from whatever region they hailed. Sidney Fisher qualified his first, critical evaluation of the Mason sisters with the reflection that genuine refinement was "met with rarely enough, anywhere."[50]

Philadelphians resolved their ambivalence about southern men in the same way. George Cadwalader, a Philadelphia gentleman who owned Maryland farms worked by slaves, clearly confused Sidney Fisher. He was "a man of the world, a man of pleasure, shrewd, practical, with much business ability, [and] no education." Some of these qualities were clearly vulgar or refined, but most were ambiguous. Business acumen provided the means to live genteelly, but it might also be the mark of the miser. A man of pleasure might be morally unfussy, or he might be debauched. Fisher still entertained doubts about Cadwalader years later, though he judged him to be without a doubt "a gentleman in his manners as he is in birth & breeding." Philadelphians found fault with their southern friends, but they praised their merits with their next breath. George Izard may have been arrogant, but he was also "a fine looking man" with "polished manners." Even the influence of slavery, so often cited as the root of southern troubles, was unclear. Finishing Frederick Law Olmsted's *Journey in the Seaboard Slave States* (1856) early in 1856, Sidney exclaimed, "I had no idea of the ignorance, poverty, & barbarism that slavery had produced among the *whites.*" Yet, during a visit to Baltimore in the summer of 1848, he had observed "a thousand evidences of the influence (good and bad) of slavery," among which he singled out a "well-bred air" among southern men, who "look as if they were accustomed to take their ease and enjoy leisure." Considering how much Sidney Fisher prized the ideal of the leisured gentleman, that was strong testimony for the merits of enslavement.[51]

The political convictions of the Fishers and their friends were every bit as reactionary as those of the Manigaults, but they lacked the will and perhaps the imagination to use social affairs to advance an ideological agenda. Both families believed that most Americans were little more than vulgarians. While in Paris in 1836, Henry Middleton Jr. even went so far as to compliment that infamous bête noire of American manners, Frances Trollope. The derision of British Grand Tourists toward their American counterparts, joked Middleton, "might serve as a sort of supplement to Mrs. Trollope's *Domestic Manners* [1832]." In this hyperpatriotic and anti-British period, such sentiments placed the Middletons on the fringes of American culture. But this generation, unlike their forebears, had resigned themselves to marginality. During the first secession crisis of 1850–51, Williams Middleton teased his brother-in-law for railing against "a wild, proscriptive, & unbridled *democracy*" instead of resisting it in the public arena. But Sidney Fisher believed that the only appropriate role for a gentry class in a democratic society was on the margins. "Our government is a democracy. Democracy is its genius, its essence, its life," he stated resignedly. Therefore, the "proper position" of the Whigs—the party of gentlemen, Fisher assumed, was "in opposition."[52]

The Fisher circle's sociability reflected their defeatism. Their social affairs embodied no end except pleasure itself. Unlike the Manigaults, the Fishers and Middletons pursued indulgence for its own sake. Eliza Fisher cut off a brief letter to her mother in order to "get ready for my walk, that I may dispell the effects of last night's dissipation." A string of late-night affairs had left her with a vicious hangover, and she had another dinner party to attend that evening. "Only think of my dissipation!" she boasted to her mother. Respectable folk would have been scandalized enough by such behavior, which violated norms of feminine propriety. But Eliza and the other women of her circle went further. American popular culture exalted motherhood as women's highest calling, but Eliza delegated most of the child-rearing responsibilities to servants so that she could pursue a life of pleasure. When she and her husband visited Newport during the summer of 1845, they left their daughters Lily and Sophy under the care of their nurse for a full month. In aristocratic fashion Eliza believed herself to be entitled to the benefits of motherhood without incurring its responsibilities. When it appeared that her baby, Mary Helen, "love[d] her nurse so much better than her," Eliza lashed out at the young woman as an "Alderney"—a breed of cow.[53]

Privileged northerners had grave doubts about the state of genteel culture in the South. The sectional crisis drew new attention to southern gentility and invested it with greater significance. If slavery made the South less cultured than the free states, would not its spread doom the West to barbarism? And was it not likely that the contagion would infect the North in time? The Fishers' Philadelphia circle was not immune from these concerns. Nor, it should be said, were the Middletons and other southerners who moved within high circles in Philadelphia society. They articulated a critique of free society that mirrored, in many of its particulars, northerners' complaints about the slave states.[54]

Nevertheless, social elites in the North and South recognized that they shared qualities that made these distinctions seem minor by comparison. Both the northern and southern gentry recognized in each other a common gentility that dampened—but did not mute—the significance of sectional differences. Their common experiences, which distinguished them from all but a small minority of Americans, drew northern and southern gentlepeople together. When Sidney Fisher met Williams Middleton for the first time in 1840, he found that he had "tastes that agree with my own."[55]

". . . we do not altogether sympathize"

Though they tried to segregate politics from sociability, the sectional conflict taxed the Fisher circle's capacity for civility. Joshua and Eliza Fisher's efforts to explain northern opinion to the Middletons were rebuffed. They became proponents of radical southern positions on the extension and merits of slavery. Sidney Fisher drifted in the other direction. As the South's demands became more strident, he grew increasingly hostile toward the region and to its northern defenders. Until the actual outbreak of hostilities, reactionary Philadelphians compartmentalized public affairs from private life. At times of extreme pressure the kettle boiled over. But until 1861 the value gentlefolk placed on their insular world enabled them to resist the centrifugal forces of sectional animosity and keep their society together, if just barely.

Sectionalism's potential to shatter the gentry's social world appeared with alarming suddenness in 1844, when Eliza and Joshua innocently proffered their opinion that the South was unwise to press for the annexation of Texas. When Eliza's mother passed on these views to her husband and sons, they lectured her "not to trouble yourself about the interests of the South which you may well imagine will be safe if left in the hands of its own children." Southerners did not need "the officious kindness" of their "Northern brethren," they informed them. In case Eliza and Joshua did not get the message, Mary told Eliza pointedly that her brothers "are equally ardent in the cause of annexation & yr. Father perfectly *red hot.*" But the Fishers proved to be more stubborn than the Middletons wished. Eliza and her husband were, to be sure, careful to avoid taking sides against the South. Indeed, they became increasingly sympathetic to southern demands regarding slavery throughout the 1850s. But they did take pains to try to cool off the hot-headed Middletons, particularly Eliza's fire-eating brother Williams, whose letters burned with screeds against "the accursed union." Joshua agonized as sectional tension threatened two of the things he cherished most—the nation and his social circle. "I do not know what part to take in the great questions that now divide our country," he wrote to his cousin John Brown Francis at the end of 1860. Secession was "utterly wrong," he believed, but neither could he "sympathize with the Puritanic fanaticism in the North."[56]

Perhaps because his ties to the South via his wife, Bet, were less direct, Sidney Fisher suffered less inner turmoil than his cousin. During the 1850s he became progressively more alienated from the South, though his bonds with southerners remained strong. Like many northerners, Fisher was shaken by the South's response to Preston Brooks's savage beating of Charles Sumner at his desk in the Senate chamber in May 1856. "The worst feature in the case," he believed, was "the diabolical manner in which the Southern papers all sustain & praise Brooks." He railed against filibustering and the controversy over the admission of Kansas as a slave state, both of which were to his mind nothing less than "southern aggression." He deplored the South's periodic threats to secede. Although Sidney did not believe that the South would carry through on these threats, their outrageousness indicated the depths to which southern political culture had fallen. "This is the old threat always made by the South when it is opposed," he declared in 1856. "Unless thro some sudden excitement & madness, they will never dissolve the Union."[57]

Precisely because their ties to like-minded southerners were so close, few Philadelphians could take secession threats seriously. Warm relations between upper-class planters and Philadelphians persisted until the very eve of hostilities. In the midst of the Kansas controversy late in 1856, Sidney attended a wine tasting at which he met William Summer, a South Carolina writer. Despite their political differences, Fisher found him to be "social, intelligent, [and] gentlemanlike." It was these bonds that allowed him compliment a pro-southern essay by Robert J. Walker of Mississippi as "the best on that side I have seen." To preserve these bonds the women and men of the Fisher circle were willing to go to great lengths. They simply stopped discussing political affairs when they threatened to disrupt civility. When a long tirade on southern rights threatened to taint a warm family letter in 1851, Williams Middleton wrote, "I will say no more now upon this subject." Joshua returned the compliment in a chatty 1855 letter when he declined "to write about politics foreign or domestic as we do not altogether sympathize." Although such choices inevitably created awkward silences—politics being a favorite topic of conversation among Americans—it was a worthwhile tradeoff. Their willingness to make such compromises allowed privileged southerners and Philadelphians to maintain warm relations until the outbreak of the Civil War.[58]

The Middletons, Fishers, and the families in their social circle possessed wealth beyond the imaginings of all but a few Americans. Riches were the foundation of the leisure class. Without them, families could not afford the travel, apparel, residences, and the innumerable other things large and small that, in the right combination, made genteel life. As the example of Sidney George Fisher demonstrates, however, wealth alone was insufficient. Although his exclamations of poverty mocked the conditions under which most Philadelphians lived, Fisher was relatively underprivileged compared to his peers. Education ranked high among the qualities that distinguished gentle from ordinary people. Refined behavior was supposed to be effortless, an outward reflection of inward grace, but it was anything but natural. For both women and men, cultivation emerged from years of study and its application in public spaces.

Mr. and Mrs. Ralph Izard (Alice Delancey), 1775 portrait by John Singleton Copley. Oil on canvas; overall 174.6 × 223.5 cm. (68 ¾ × 88 in.); framed 203.2 × 254 × 10.2 cm. (980 × 100 × 4 in.). Museum of Fine Arts, Boston; Edward Ingersoll Brown Fund, 03.1033. Photograph ©2006 Museum of Fine Arts, Boston

"China Retreat, Pennsylvania: The Country Seat of Mr. Manigault," from William Russell Birch, The Country Seats of the United States of North America: With Some Scenes Connected with Them *(Springland, near Bristol, Penn.: designed and published by William Birch, enamel painter, c. 1808–09). Courtesy of the Winterthur Library, Printed Book and Periodical Collection*

Sidney George Fisher (standing on left). Historical Society of Pennsylvania

Eliza Middleton Fisher. Historical Society of Pennsylvania

Joshua Francis Fisher. Historical Society of Pennsylvania

Williams Middleton. Historical Society of Pennsylvania

Mademoiselle Adèle Sigoigne, portrait by Thomas Sully. Courtesy of the Huntington Library, Art Collections and Botanical Gardens, San Marino, California

"Departure for a Boarding School" and "Return from a Boarding School" by John Lewis Krimmel. Courtesy of the Winterthur Library, Joseph Downs Collection of Manuscripts and Printed Ephemera. These early-nineteenth-century images illustrate the viewpoint of critics of gentility, who considered boarding schools aristocratic, impractical, and un-American

The President's House on the Second Campus (1802–28) of the University of Pennsylvania. Collections of the University of Pennsylvania Archives

The Medical Department building on Ninth Street Campus (c. 1850) of the University of Pennsylvania. Collections of the University of Pennsylvania Archives

John Vaughan, portrait by Thomas Sully. Courtesy of the American Philosophical Society

View of Chestnut St., *by J. T. Bowen. Courtesy of the American Philosophical Society*

State House, *by A. Kollner. Courtesy of the American Philosophical Society*

The Penitentiary, Phila., *by W. Walker. Courtesy of the American Philosophical Society*

View from the State House (looking east), *by John Caspar Wild. Courtesy of the American Philosophical Society*

Fairmount Gardens with the Schuylkill Bridge, *W. H. Bartlett. Courtesy of the American Philosophical Society*

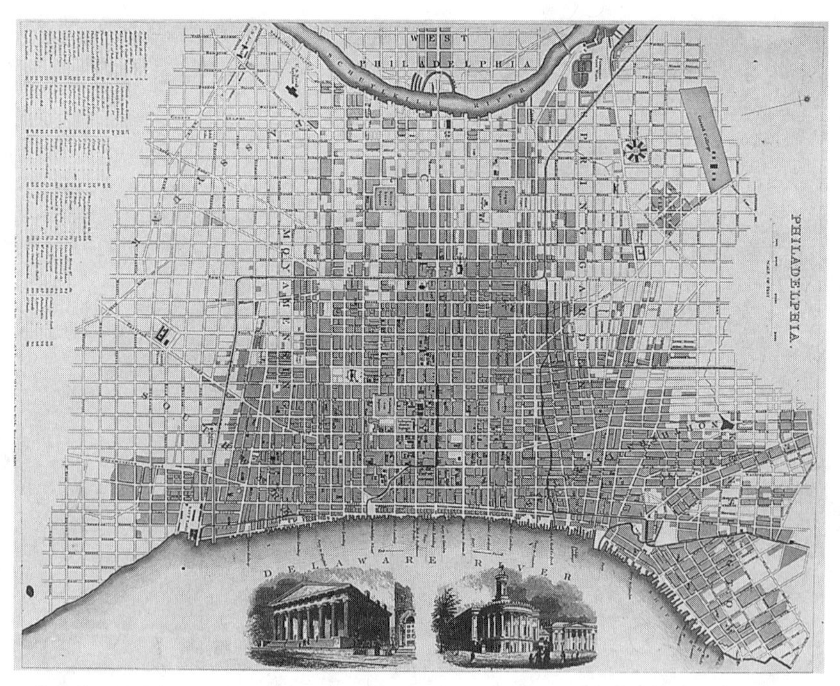

Map of Philadelphia, September 1840, showing locations of the Fairmount Water Works and the Eastern State Penitentiary in the northwest portion. Society for the Diffusion of Useful Knowledge. Courtesy of the Library Company of Philadelphia

3

"Your appropriate sphere as a lady"

Philadelphia's French Schools

Leading society women such as Elizabeth Middleton Fisher were made, not born. Wealth and family name were necessary conditions for social prominence, but they were not sufficient. To shine in select companies, a gentlewoman needed a variety of skills. She had to master a variety of complex dance steps and keep up with new ones. Proficiency at the guitar, piano, or harp enabled her to entertain company. Fluency in French served as a marker of privileged status, but it also had practical value if one traveled abroad. Knowledge of current events, literature, history, and even science was necessary for women to engage in the quintessential social activity, conversation. Everyone agreed that the possession of these skills was desirable, even necessary. There was some disagreement over how they might be acquired. During the first half of the nineteenth century fashionable boarding schools—often called "French schools" because of the emphasis they placed on French language and manners—became increasingly popular with wealthy, conservative families. There were dissenters. Alice Izard believed that young women needed stricter moral supervision than headmistresses were willing to provide. Even advocates of French schools feared their moral influence, to which they added anxieties about distance, expense, and practicality. Nevertheless, they flourished in every American city in the antebellum period.[1]

The popularity of these schools with privileged families underscores the depths of social-class divisions in nineteenth-century America. Conservatives' patronage of such schools illuminates their desire to separate themselves from those they considered to be their social inferiors. But it also displays the depth of values that divided privileged from ordinary Americans. Middling women and men sought to match American political independence from Europe with a distinctive national culture, but French schools continued to bow toward the Old World. And while respectable people praised the virtues of democracy, conservative boarding schools embraced elitism. These differences were the root of an uneasy exchange between Langdon Cheves, the South Carolina planter who served as president of the Second Bank of the United States from 1819 to 1822, and his business agent, Ainsley Hall, who asked Cheves to recommend a Philadelphia boarding school for his niece in 1819. Hall requested Cheves to exclude "all French Characters" from consideration, being opposed to "the fashion of introducing French manners & customs, to the observations of young ladies." Cheves suffered considerable difficulty in complying with this request, lamely recommending that he look into the Moravian academy in Bethlehem, Pennsylvania. Though Cheves could not in good taste spell it out, a social chasm separated the lowcountry patrician from the upcountry businessman. While he lived in Philadelphia, Cheves sent his daughters to the school of Madame and Monsieur Picot, Gallic spirits from St. Domingue with both feet planted firmly in the ancien régime.[2]

French schools were hardly the only type of educational institution for women to flourish in this period. It is no exaggeration to say that Americans at this time witnessed an explosion of educational opportunities for young women. These took a variety of forms, reflecting their widely divergent purposes and constituencies. After the Revolution reformers such as Judith Sargent Murray and Benjamin Rush argued that colonial educational practices—whereby the few women who received anything beyond a basic education did so under a tutor or at a boarding school that gave short shrift to academic study—were inappropriate for a republic. The United States could neither tolerate ignorant women nor those reared according to the standards of aristocratic Europe. Rush put his ideas into practice at the Young Ladies' Academy of Philadelphia, which he helped found. The school taught its young charges to write legibly, speak clearly, keep accounts, master geography and the natural sciences, and practice vocal music. Rush sought to produce wives who would be helpmates and companions, not ornaments. And since the republic required an enlightened electorate, educated ladies would be able "to concur in instructing their sons in the principles of liberty and government."[3] Despite the new enthusiasm support for women's education remained limited. Critics contended that resources were wasted on them. Women should become wives and mothers, not public figures. Parsimonious state and local governments provided very limited support to female education, which remained largely in private hands in the early decades of the century.[4]

Matters improved after 1830, when public funding increased, a number of women's colleges opened, and private academies flourished. The main impetus for these developments was the rise of a self-conscious middle class that saw education as the foundation of upward mobility, Christian piety, and economic independence. Women benefited directly from these developments. Expanding on early national concepts of the republican citizen and republican mother, antebellum educational reformers such as Catharine Beecher, Lydia Maria Child, and Emma Willard argued that women, as the moral guardians of American society, needed instruction to raise children, manage households, and engage in social reform. The more radical among them, including Willard and Elizabeth Blackwell, built on Judith Sargent Murray's writings to suggest that education would provide women with options beyond marriage and motherhood on the one hand and spinsterhood and poverty on the other. Their writings, as well as the efforts of educational advocates such as Horace Mann, led to the founding of public-school systems for boys and girls, particularly in the New England states.[5]

The establishment of private academies also intensified. These schools focused on inculcating a middle-class ideology of "frugality, temperance, and economy," as Abigail Mott, a leading educational authority, declared. They eschewed subjects such as watercolor, voice, and dancing in favor of mathematics, science, and history. State-funded public schools focused on basic educational goals, but private academies such as the Patapsco Female Institute, led by Willard's younger sister Almira Lincoln Phelps, were more ambitious. They promoted an ideology of middle-class women's activism. They urged women to use their moral influence to improve society, which in turn led some of their graduates to agitate for women's rights. Though most of these schools—particularly publicly funded systems—were founded in the North, the South saw its share of private academies, and the region led in the development of women's colleges.[6]

The French schools attended by young southern women in Philadelphia could not have been more different from these institutions. They were similar in many ways to the schools that became a "minor industry" in colonial seaboard cities, catering to the daughters of prominent merchants and planters. They were academically more rigorous. Most offered instruction in history, geography, natural science, natural philosophy, and other academic subjects also found at boys' schools. But they also stressed the so-called ornamental branches of education, such as drawing, painting, and voice. More important, French schools followed their colonial forebears in orienting themselves toward the Old World instead of the New. American cultural leaders devoted themselves to establishing a national literature free of slavish imitation to European models. One writer in the *North American Review* ascribed America's lack of "intellectual vigour and originality" (given unintentional credence via the British spelling) to its "enslaving" dependence on English culture. French schools' curriculum and student culture were thoroughly aristocratic, endorsing elitism and the primacy of European culture at the very time when educational reformers exalted education as the primary engine for national development.[7]

Conservative women's academies not only oriented themselves squarely toward the Old World, but embraced the ethic of its reactionary upper classes as well. French schools repudiated the egalitarian trends of early nineteenth-century American education. Their expense defied a growing consensus that education ought to be accessible to rich and poor alike. The expense of a boarding-school education put it beyond the reach of all but the wealthiest Americans. The costs included tuition, room, board, and a host of other charges. When Eliza Spragins entered Anne Marie Sigoigne's academy in 1839, her father handed over $425 for a semester's worth of tuition, room, board, and washing. Two years before, Isabelle Hunt's parents received a bill for $950 for a full year at the school. Besides the usual charges, Hunt's bill itemized fees for additional lessons in piano, voice, and guitar as well as charges for concert tickets. Deborah Grelaud charged $500 per year in 1839, a year when skilled workers in the American Northeast earned about $1.50 per day. These costs rendered French schools a preserve of the privileged. The mere fact of attendance there announced a family's aristocratic aspirations. When Nathaniel Ware of Natchez placed his stepdaughter Mary Jane Ellis at Mrs. Phillip's school in 1820, his business partner agreed that "with Mary Ellis's fortune, her education and the proper formation of her mind to fit her for society and happiness in mature life is of much more importance than a few thousand dollars."[8]

Little wonder then that leaders such as Mann and Almira Phelps bitterly criticized French schools as reactionary and un-American. It was not merely that Mann and Phelps were repelled by these values. Rather they feared their influence in American society, whose commitment to middle-class values seemed fragile and provisional. The persistent allure of European aristocracy was everywhere evident, from travelers' eagerness to be presented at European courts to the reading public's obsession with the opinions of travelers from across the Atlantic. Basil Hall, a Scottish naval officer who toured the United States in 1827–28, marveled at "the anxiety with which the opinions of a foreigner were sought for with regard to many insignificant topics, upon which his sentiments might have been thought very little."[9]

Critics heaped derision on French schools precisely because they recognized the threat they posed to the vision of a respectable, middle-class republic. Almira Phelps charged that boarding schools turned virtuous girls into "artificial creatures, made up of artificial looks and smiles." A New England writer charged that at conservative academies, "little attention is paid to any thing but fashion and folly." Pupils learned little of use, instead "becom[ing] adepts in dress and other frivolities." Yet this writer recognized the "modish" allure of such schools for the great many Americans still under the thrall of aristocracy. Middle-class reformers feared French schools because they recognized that reactionary Americans still exercised considerable influence on national life. They were right, moreover, about the reactionary character of these schools. "Elite parents'" patronage of French academies was "a clear sign of class cohesion," observed Steven Stowe. Their expense, exclusiveness, and curriculum provided "evidence of elite position and power." Middling people were as aware of the character of these schools as they were of the precariousness of the republican experiment. All their work in building a nation free from the political and cultural authority of Europe might still be undone.[10]

". . . the favored spot for female education"

Southerners did not have to send their daughters to Philadelphia to be educated in the European manner. The South had a number of elite boarding schools of its own. St. Mary's in Raliegh and the Salem Academy in Salem, North Carolina, the Montpelier Institute in Savannah, and Madame Talvande's school in Charleston, attended by the famous diarist Mary Chesnut, were nearly indistinguishable from Philadelphia schools. In the spring of 1821 Thomas Percy, his wife, and young son traveled with several friends overland from Huntsville, Alabama, to Philadelphia to install his niece and several other girls in an academy there. The journey, which Percy described as "laborious & slavish," likely deterred many planters from considering Philadelphia. Moreover, some southerners feared that Philadelphia's politics might be unsuitable. An Alabama planter touring the North in 1826 expressed a fear that "girls here are imbibing habits and manners not perfectly congenial with those of the people of the South."[11] These obstacles ensured that most planter-class women who attended French schools did so in the South.

However, Philadelphia was a popular destination among cosmopolitan southerners. It was favored over many northern cities because of its proximity to the South as well as for the southern orientation of its upper crust. Emma and Annie Shannon, pupils at St. Mary's Academy in Burlington, New Jersey, just across the Delaware from Philadelphia, gleefully confirmed these attitudes in an 1857 letter to their mother. "You can't think in what estimation Southerners are held here," Emma exclaimed. "They are looked up to as superior beings." The city enjoyed several advantages over southern towns and cities. It was the leading destination for French exiles fleeing revolutions at home and in St. Domingue in the late eighteenth century. When poverty threatened, some of these figures established schools to tap into the American demand for European-style educations. They included Anne Marie Sigoigne, Deborah Grelaud, Charles and Marie Picot, who had fled the Caribbean, and Marie Rivardi, chased out of France by the Terror.[12]

Rivardi's experience was typical. An habitué of the Hapsburg court, she exhausted her savings just a few years after reaching the United States. Running a school was hardly the ideal option for such a privileged soul, but it was better than poverty. Moreover, it did not require her to master any new skills or to submit humiliatingly to employment. On the contrary all that was expected of her was to pass on what she already knew. Conservative Americans "were anxious for their children to be exposed to the sophisticated culture of Europe," and these schools delivered that at far less cost than the alternative, a trip across the Atlantic to tour the palaces and museums of the Old World.[13] Other American cities boasted French schools, but none other possessed the concentrated population of exiles and sheer number of institutions that Philadelphia could offer. Susan Petigru King fictionalized the process by which her parents placed her at Madame Guillon's school in Philadelphia after several years at a Charleston academy in her novel *Lily.* After visiting schools in a variety of cities "in every direction," Alicia Clarendon's parents "came to the conclusion that the city of straight streets and very clean bricks was the favored spot for female education."[14]

Planter-class parents also installed their daughters in Philadelphia schools because they wanted to expose them to a more cosmopolitan atmosphere than southern schools could offer. In general adults believed that the South lacked the educational and cultural resources to produce urbane young women. Samuel Brown, a planter who had studied under Benjamin Rush at the University of Pennsylvania, feared the influence of the "mental inactivity" of the South on his adolescent daughter Susan. Following the death of his wife, he resolved to educate his children "in some part of the world where Religion morals & social order were better established" than in the rough frontier communities of Natchez, Mississippi, and Lexington, Kentucky, where he owned plantations. New Orleans, the home of several reputable schools, was nearby. Moreover, his brother James, a future U.S. minister to France, lived there. However, Brown ruled out New Orleans because of its "profligate frivolous society." He decided that in Philadelphia Susan would acquire "the principles of a better taste & more delicate manners than those which prevail" in the South. He soon enrolled her in the celebrated academy of Mrs. Phillips, with whom he was so pleased that several of his friends enrolled their daughters there on the strength of his recommendation.[15]

As Brown's musings indicate, Philadelphia academies were likely to appeal to southerners with preexisting links with the North or cosmopolites who craved contact with the world outside the South. To urbane planters Philadelphia was a far more familiar place than many regions of the South. When Martha Richardson, a Savannah gentlewoman, visited some mineral springs in Green County, Georgia, during the summer of 1821, she wrote that not only was that region of the state "a new world to me," but so were "the customs & habits of the people." Travel to the North was "a passion" among her Savannah circle, she informed her nephew James Screven the following summer. One young man had departed for Princeton, and still another for Yale, while Georgia Bryan, a cousin, had left "to be placed at a boarding school in Philadelphia."[16] Most southern women, including the majority of slaveholders, were "destined to live their lives in the relative isolation of the rural South," but a considerable number were urban dwellers for at least part of the year. Harriet Horry Ravenel remarked that the great planters of South Carolina were "as much town folk as country gentlemen," and the same was true of wealthy slave owners throughout the region. Even rural planters recognized the merits of urban life, with its concentration of cultural amenities, social activities, and shopping. In early 1819 John Williams Walker, an Alabama planter and Princeton graduate about to serve a term in the U.S. Senate, anticipated visiting his friends in the "Quaker City," where he planned to enroll his daughter Mary Jane at school and let his wife "figure away among the great."[17]

Because they were likely to be well-traveled, many planters enjoyed personal connections with Philadelphians. Southerners established relationships with northerners in a myriad of other ways as well. Business engagements, memberships in scientific societies, and political activities coordinated through correspondence all created bonds between the sections. These links made Philadelphia a primary destination for southerners seeking a finishing school for their daughters. Few parents were willing to separate themselves from their children for months, and perhaps years, without some arrangement for their security and well-being. They trusted school administrators of course, but, as one parent understood, a "great boarding school" could devolve into a "heartless community" to an insecure young woman. The best arrangement was to employ friends or relations to provide boarders with a home away from home. Locals monitored girls' health, well being, financial condition, and social development for their southern friends and relations. They invited them to their homes for meals and social events, took them shopping, and integrated them into their family rituals. These links were a powerful inducement for southern parents. Parents admonished girls to write frequently and candidly about their progress. William Gaston instructed his daughter Susan to write so frequently and in such detail that he should be able to "accompany you in my mind through the occupations of the day." But young women were often maddingly laconic. "Mother really I do not know [what] to tell you," Appie King told her mother at the end of a brief 1852 letter, "for there is nothing here which will interest you." Southern parents depended on their Philadelphia connections for more candid appraisals of their girls' progress than the daughters were disposed to provide.[18]

Philadelphians proved to be indispensable to southerners at every stage of the boarding-school process. Academies advertised in southern newspapers and information about them spread via word of mouth, but Philadelphians provided on-the-spot, authoritative information on the relative merits of the city's schools. When Annie Iredell, the daughter of North Carolina planter James Iredell Jr., came of age, her father asked James Mease and his sister Isabella for their opinions about the school run by Charles and Marie Picot, French exiles whose academy was popular with North Carolina families. The Picots were "devoted to their pupils," Isabella assured Iredell. Their qualifications were "solid & in some instances astonishing." A boarding-school education entailed considerable expense—six months at the Picots' school cost $230, including extra charges for laundry, board, and piano and voice lessons—and parents needed assurance that their money would be well spent. Philadelphians provided these services as acts of friendship, but they also benefited in tangible ways. Being invested with such a weighty responsibility signified the esteem in which they were held by their peers in the upper ranks of southern society. Isabella assured Iredell that to "render [Annie's] residence there as comfortable as possible" would be a "pleasure," not a duty. In fact, Mease personally selected Annie Iredell's voice instructor and dined with Annie at her brother's house every Saturday. James Mease also wished to impress on the Iredells the extent to which he took his responsibilities to Annie seriously. When she was under his roof he treated her as "*my* daughter," he assured his friend, and when she visited the nearby home of Dr. Nathaniel Chapman, "she is almost at home as with us."[19]

Both southerners and their Philadelphia hosts appreciated the gravity of guardianship. Monitoring a young woman—and sometimes two or three when sisters traveled to school together—entailed much more than dropping by every few weeks. Because sending a girl to Philadelphia meant withdrawing from her day-to-day life for a year or more, southerners expected their local contacts to accept the responsibilities of parenthood. In the late summer of 1823 William Gaston returned to New Bern, North Carolina, after leaving his daughter Susan in Philadelphia. He appealed to Joseph Hopkinson's "friendship" to "from time to time" examine her progress. Moreover, he empowered his friend to "direct such changes and give orders for such helps as a Father on the spot would deem himself authorized to make." Guardianship always grew out of a preexisting relationship, but it renewed and strengthened that bond. Welcoming a friend's daughter into one's household strengthened links between northern and southern elite families, extending intersectional links into a new generation.[20]

As the language of parenthood testifies, this practice established links of fictive kinship between Philadelphians and southerners. Elizabeth Bordley and Eleanor Parke Lewis had boarded together at a Philadelphia school in the 1790s. When Lewis in turn sent her daughter there to attend Deborah Grelaud's school in early 1815, she engaged her friend to accept the responsibilities of motherhood—a role Bordley readily accepted. She pledged to protect "*our* daughter" from the "Syren voice" of the big city while she was away from Virginia. Parke Lewis thrived at Grelaud's school, and while she enjoyed the doting hospitality and access to Philadelphia's society provided by her mother's friend, she did not require more serious supervision. The same was not true of her sister Eleanor, who died while enrolled in Grelaud's school at the age of fifteen in 1820 because, her mother believed, the Frenchwoman had been tardy in soliciting "proper attention & medical advice." Selecting a Philadelphia guardian for one's children was a serious business.[21]

The responsibilities of fictive parenthood were mundane but essential. Managing youngsters' money was a common task. When Susan Shelby lost all her cash a few days after arriving at the Picots' academy—she believed the servants had pilfered it—her mother told her to "call upon Mr. White stating that you have been robbed" to seek an advance, which was supplied. Equally important, guardians monitored women's educational progress, making sure they minded their studies and were receiving the proper instruction. Very often parents supplemented the boarding-school curriculum when they wanted their children to learn a skill not offered at the school or when they found the staff wanting. The latter was the case when Mary Wiley, a student from Macon, Georgia, desired voice lessons. Mary Gill, her headmistress, wrote her parents that Wiley's New York uncle had "authorized us to engage Herelle (probably the most skillful master in the country) to give her singing lessons." Gill pledged to retain the singing master "with as little delay as possible." Gill was not seeking the girls' parents' permission to do so. As a courtesy she was informing them of a decision made by a nearby relation who enjoyed the authority to act in a young woman's best interests, which in this case entailed spending a considerable amount of her parents' money.[22]

The eagerness with which Philadelphians met their responsibilities testifies to the strong bonds that linked together privileged northerners and southerners. Emma and Annie Shannon, daughters of *Vicksburg Whig* publisher Marmaduke Shannon, benefited from their father's numerous Philadelphia contacts while they boarded at St. Mary's Academy in nearby Burlington, New Jersey. Mississippians passing through the city provided the girls with newspapers from home as well as with issues of *DeBow's Review.* Locals also had them over at Christmas, when the girls would have felt their distance from Mississippi most keenly. Unwilling for them "to be without something whilst among others that received presents," their father arranged for his friend William Stuckey to deliver packages of apples, dried beef, fruitcake, crackers, and pickles in time for the holiday. Though Shannon merely requested that Stuckey buy and send off the goods, the Philadelphian resolved to deliver them in person so that he might give their father his personal impressions. Stuckey's reasons epitomize the strength of the bonds of kinship and friendship that prompted many southern mothers and fathers to entrust their daughters to Philadelphia boarding schools. "We all feel so much anxiety to see them & know how they do," Stuckey assured his friend. "I have had children," he empathized, "and I know what a Parent's feelings are."[23]

". . . atmospheric pressure, moral and religious"

Though most students enjoyed the company of relations, friends, or guardians from the local area, the boundaries of the community of fellow scholars and teachers defined their day-to-day experience and exercised the most influence on their development. The boarding-school world brought together young women from disparate parts of the nation, along with international students, and forged them into a close-knit sisterhood whose existence often persisted well beyond the time they returned home. Both schools and the experiences of the young women who lived there differed widely. Classic French schools such as those run by the Picots or Deborah Grelaud enrolled only a handful of students. Most of them lived in the house, though academies might also enroll some day students from the city. Thus, young women's academies established a relatively closed world where students could develop into gentlewomen with little interference from the outside world. As Reverend George Washington Doane, rector of Burlington's St. Mary's, told parents at the school's 1849 commencement, "The child goes to school, to study, and be trained. The training is by atmospheric pressure, moral and religious. To let it up is to lose its influence." While Doane's characterization of "applications of permission for a child to go home, for a day or two; or to visit a friend; or to repair, on Saturday, to the city" as "positive evils . . . destructive of good order" was extreme, all schools sought to construct a total environment where girls could be molded into refined young ladies.[24]

Parents, faculty, and students had different understandings of the importance of constructing a closed world. Doane understood that parents would resist relinquishing control over their children. "Distance itself will raise its doubts," he warned. Headmistresses confronted these fears with assurances that their institutions tempered academic rigor with familial love. They told parents that the school community was itself a family. They also insisted that girls write home frequently.[25] Mary Gill strove to make this clear to Mary Wiley's wavering parents in 1853. During the daytime Mary took lessons in moral philosophy, French, ancient and modern history, and "the Analogy of Religion to Nature"—the last an unobjectionable approach to religion in nondenominational French schools. Occasional exercises in penmanship, spelling, geography, and grammar rounded out her week. Dinner followed the end of classes at two P.M., after which the girls took a walk together. Afternoon tea was crammed in between music lessons and a private study period. Evenings were spent "in the parlor with Sister Sarah [to] mend their clothes while she reads or talks to them." Lights went out at nine P.M. "By these regular arrangements a great deal can be accomplished," Gill explained. Besides, it was believed that young people needed restraint and discipline lest they "keep [themselves] in a continuous excitement."[26] But Gill took pains to explain that Mary Wiley received more than structure at her school; she was also loved as a daughter. Mary had become a member "of our family," she assured her parents. "In promoting the welfare of your dear daughter we have a great and common interest and one that requires cordial cooperation." Students, not surprisingly, had a different opinion than their parents and teachers about the benefits of this regimen. "I don't believe we will learn," declared Emma Shannon after detailing her busy weekly schedule to her mother. "I don't think my constitution, for one, can support" such "unremitting study."[27]

The Shannons soon overcame their reservations about St. Mary's Hall. A key factor in their change of mind was the support of their fellow students, particularly the large contingent of "Vicksburgers" and other southerners, who formed a community amid the larger population of northern students. At the Burlington academy, Annie Shannon explained, southerners "form a set from which the northerners are excluded." Boarding-school students formed groups around a variety of interests in order to accommodate to the anonymity of institutional life; "timid sensitive characters," one observer noted, found it difficult to adapt. Southern students naturally gravitated toward one another, particularly during their first days at the academy. Most did so without the ideological self-consciousness displayed by the Shannons who, their rhetoric aside, made close friendships with northern students.[28] Young women befriended each other because their southern roots mollified homesickness, introducing an element of familiarity into an alien environment. Usually older students befriended newcomers, easing their transition into the urban, boarding-school world, introducing them to its rituals and unwritten rules, and supplying confidential information about classmates. Students usually went on to make close friendships with girls their own age regardless of their point of origin, though southerners identified themselves as a community-within-a-community.

As Emma Shannon's description illustrates, most friendships between planter-class girls were not expressions of regional solidarity. They were extensions of preexisting family or neighborhood connections. Because French schools tended to attract pupils via word of mouth, they often catered to particular webs of family or locality in the southern states. These networks could be important sources of income to Philadelphia schools. When Susan Polk's parents withdrew her from the Smith sisters' school in 1835, Helen Smith told Polk she feared the action would "injure their reputation with strangers." Sister followed sister, cousins attended school together, and families recommended schools to each other. The result was a network of kinship and friendship that softened young women's transition from plantation to classroom. When Jennie Ellis entered Sigoigne's school in 1837, she immediately sought out her six fellow Virginians. "Clannish as ever we are generally together," she told her brother of her early days at the academy. Margaret Mordecai felt butterflies as her eyes fell on the school run by the Hawks sisters, where her sister Ellen had been educated. "You did not know how much my heart did beat when we drove up to the door," she admitted. But a familiar face soon appeared to assuage her fears. "Susan Polk was the first person I saw she carried me up into her room which is the same as when you were here," she told her sister. Polk, a family friend, helped Mordecai feel at home at her new school. In several days she was an intimate member of the boarding-school community. "You have no idea of how kind every body has been to me," she reported to her sister. "All of your old friends received me as kindly as they had known me all of their lives."[29]

Parents sent their daughters to Philadelphia schools so they might return to the South as elegant, polished gentlewomen. They did not fear that girls' friendships with their schoolmates might draw them away from the South. As urbane Americans, they saw little to fear in intersectional friendships and much to admire. They also interpreted their girls' acceptance by their boarding-school community as a sign of their developing social and academic skills. Moreover, status-conscious planters saw their daughters' intimacy with the daughters of other elite families as a confirmation of their social rank. William Gaston assured his daughter Susan that he approved of her acquaintance with a Pennsylvania student, for "to be honoured with her friendship is no small proof that my child deserves the affection of the good." Youngsters were hardly insensitive to questions of status themselves. They were drawn to women from prominent families or those who mastered the genteel arts. When the parents of Hester Van Bibber of Matthews County, Virginia, withdrew her from school during the 1816 summer recess, her fellow scholars bombarded her with letters expressing their "surprise" and "disappointment." Van Bibber's friends, such the day scholar Elizabeth Buchanan, would miss her company, but they also mourned the loss of the status that they enjoyed merely by association with their friend. "How cruel in you to leave us thus," Buchanan complained. "Besides the loss of your society *out* of school, what shall we do without you in?" As her last comment suggests, Van Bibber excelled in the classroom as well as the parlor. Her mastery of French, the ultimate "ornamental" subject, rendered her loss a serious one. "I must now hobble through my French book all alone!" Buchanan cried.[30]

As Elizabeth Buchanan's anguish reveals, young women developed extraordinarily close friendships during their boarding-school years.[31] Intense likes and dislikes were inevitable products of the close quarters of the academy world. Students attended classes, dined, studied, socialized, and relaxed as a group. Small groups of girls became even more intimate, as friends shared bedrooms and even slept three or four to a bed. An 1842 evening during which Sally Clay and two friends "laughed talked nearly all night about different things" as they "slept in *one bed*" brought back to mind the time when they and their former roommate Lizzie Spragins of Halifax, Virginia, were "all there together" at "*the old Madam's,*" Sigoigne's school. The sleeping arrangements, Clay remarked to Spragins, "really reminded us of old times." These close contacts and the relationships they produced were not incidental consequences of the boarding-school world. As both its advocates and critics recognized, they were its very essence. Personal relationships forged young women into a leisure class. The reactionary atmosphere of the faculty and curriculum, together with the conformist pressures of the student body, encouraged girls to see beyond regional, religious, ethnic, and even national differences to recognize their common values and interests.[32]

Fanny Fern, an outspoken middle-class writer, saw clearly the ideological implications of boarding schools' living arrangements. Mothers should be aware, she warned, that "the distant home of the her daughter's room-mates is located within the charmed limits of fashion; that a carriage with liveried servants (that disgusting libel on republicanism), stands daily before their door; that the dresses of these room-mates are made in the latest style, and their wrists and ears decked in gold and precious stones." Upper-class mothers were quite aware that the social structure of French schools inculcated reactionary principles; that was the point of sending girls away. Fern recognized that the aristocratic allure of boarding schools appealed to middle-class Americans, and she perceived the threat to republicanism that this attraction posed. She was hardly alone in making this observation; by the 1820s, as Richard Bushman has observed, "castigation of female boarding schools was already a stock critique of gentility." These critics saw that middle-class patronage of boarding schools, like its fascination with royalty, was not a harmless conceit but a threat to republican principles.[33]

Republicanism did not only come under attack at supper tables and in bedroom conversations; the curriculum of French schools also was reactionary. Critics derided French schools' emphasis on the "ornamental" branches of education as elitist and impractical. But those skills were highly useful to privileged young women. They testified to a woman's refinement, made her an effective hostess, and helped attract a desirable husband. A wife and mother, lectured one father to his young daughter, "should be practically acquainted with domestic occupations." But a gentlewoman aspired to a higher calling: she should be "sensible," which required education. To ascend to the highest plane of all, that of the "fine lady," a woman had to master the accomplishments. A teacher at a French school urged a former pupil to carry on her studies at home so she might "acquire solid knowledge and enlarge your views" in order to "mature your taste and make yourself easy and graceful in the conversation of polished society." In vying for pupils French schools tried to demonstrate their qualifications in all of these spheres. When R. S. Roberts advertised his seminary in 1843, his list of fifteen local luminaries as references testified to its social qualities, and its "French Department" spoke to the central place of the ornamental branches. But he also boasted of its academic features, as manifested by the "Maps, Globes, a Cabinet of Minerals, Philosophical Apparatus, and a Library." Critical voices to the contrary, academics was serious business at French schools.[34]

French schools offered instruction in a wide variety of academic subjects, from history, science, and moral philosophy to belles lettres. Students attended classes where they heard lectures delivered by teachers or guest speakers. Instructors also assigned books on which students wrote essays and delivered recitations. Many French schools offered a flexible curriculum that combined assigned books and subjects with directives from parents. Richard D. Arnold of Savannah arranged for his daughter Ellen to study Latin when she boarded at a Philadelphia school in the late 1840s. He had to make special arrangements for her to do so, since the subject was not offered and was, in fact, considered so unfeminine that she became the object of ridicule. "I would not laugh at you" Arnold assured her, adding helpfully that Latin "facilitates the acquisition of all the modern languages." He also urged her to supplement her class work with especially valuable works, such as the *Spectator* and histories, which "enable you to understand the allusions and references with which you continually meet in reading." But Arnold wanted to complement the established curriculum, not replace it. He conceded that Ellen's "principal attention shall be devoted to your elementary subjects." Parents and students alike placed great emphasis on the study of literature, ethics, and history. Their enjoyment provided solace for women, who spent much time alone while men saw to business. A commencement speaker at Marie Rivardi's academy claimed that education provided women with "amusement in solitude and consolation in the hour of neglect and grief." French schools always touted the social advantages of learning above its private merits, however. As Richard D. Arnold explained to Ellen, familiarity with books and ideas turned unrefined young ladies into "accomplished and ready" gentlewomen.[35]

For the same reasons schools offered instruction in botany, conchology (the study of shells), astronomy, and other sciences. St. Mary's Hall promised "full courses of lectures in Natural Philosophy and Chemistry, with a complete apparatus, and also in Botany." Not everyone agreed all this was worthwhile. When Mary Polk informed her father of her pursuit of a variety of scientific subjects, he intervened. "The study of Philosophy, astronomy, & chemistry by young ladies," he declared, "is nothing more than an idle waste of time." William Polk insisted that Mary "fill the space they would occupy, in the study of grammar, geography, Arithmetic . . . & historical reading, with French." He instructed her to "make my wishes known to Mrs. Mallon, so that she may so arrange your studies for the future." Few parents dismissed the sciences so cavalierly. They expected that their studies would help their daughters shine in salons and ballrooms. This required at least a conversational knowledge of the sciences. When Mary Wiley returned to the South after studying at the Gill sisters' academy in Philadelphia, her teachers recommended works that combined scientific rigor with feminine delicacy, such as "Nichol's *Architecture of the Heavens* . . . Dick's *Celestial Scenery* and *Sidereal Heavens* . . . [and] Hunt's *Poetry of Science,*" though she also suggested well-regarded general works such as "Gould and Agassiz's *Zoology.*"[36]

Young women studied the sciences because gentlewomen were expected to be conversant with contemporary intellectual currents, not because they might use what they learned in the household or workplace. As one writer maintained, a "refined, an elegant, a[nd] tasteful" social circle required a company able to converse on a diversity of topics. But middle-class schools placed considerable emphasis on the sciences as well, and respectable standards of sociability also recognized the value of conversation. What truly distinguished the curriculum of French schools was their emphasis on what critics and patrons alike called the "ornamental" branches of women's education. They included training in instrumental music, usually the piano but occasionally the harp or guitar. Some women also took voice lessons. Schools often offered instruction in painting, drawing, and dance. And nearly all young women studied at least one European language. In addition to French, students might learn Italian, Spanish, and (very occasionally) German. Critics ridiculed these subjects for their superficiality, but parents, teachers, and students took them very seriously. In 1839 Julia Watson, a student at the Picots' school from Hawkwood Plantation in Louisa County, Virginia, complained to her aunt that while she had improved at the piano, her skills had not progressed "as much as will compensate me for all the practicing I do. I give every minute I can to the piano, and play one line over and over again." Watson was not angling to free herself from an onerous responsibility. To the contrary her lack of progress in a skill she valued deeply was a source of real frustration. "When I come home I shall practice still more than I do now," she pledged.[37]

Some students pursued ornamental subjects to the exclusion of all others. Jennie Ellis told her brother that "French and music you know are my chief studies, and indeed I may almost say my only ones." Such single-mindedness was the exception, however. When she discussed her schedule at Sigoigne's school for her father late in 1822, Susan Gaston illustrated how French schools integrated genteel subjects into the conventional academic curriculum. Her weekly regimen included "Rhetoric, Scientific Dialogues, Spelling . . . arithmetic, drawing, singing, harp, and piano. Writing lessons on Tuesday, history, geography, dancing, astronomy, and painting. Music I have every day two to three hours."[38] Though few schools replicated this breathless schedule, the spectrum of subjects the young North Carolinian studied was typical fare for boarding-school girls. Artistic pursuits such as painting and drawing were discretionary, but most schools insisted that students take voice and instrument lessons. Educational reformers disapproved of musical study because of its expense and inutility. Privileged Americans recognized these obstacles, which only raised musical skill's esteem in their eyes. "I am gratified that you take so deep an interest in your music," a Georgian wrote her cousin, a pupil at the same Philadelphia school she had attended years before. "The pleasure that it will afford you in years to come will fully repay you for the trouble of acquiring" it. Ornamental studies such as dancing and voice fitted aristocratic temperaments particularly well because their possession signified a natural endowment that might be heightened, but never awakened, by proper training. Singing was "an exquisite accomplishment," this Georgian maintained, "or should I say gift of nature, for cultivation never *endows* a voice; it only *cultivates,* when possessed."[39]

The study of the French language was by far the most highly valued ornamental subject. Some parents evaluated schools primarily on their ability to instruct young women in the language. In 1815 Rosalie Calvert of Maryland sent her daughter to Deborah Grelaud's French-speaking academy "after try[ing] two different tutors" over a two-year period during which her children "could not learn French." By 1817 she declared herself "quite satisfied" with her daughters' mastery of the language. As Calvert suggested, many parents preferred to educate their children at home with tutors but found the more intense environment of the boarding school to be more efficacious. Other southerners opted for French schools because they had difficulty inducing qualified teachers to venture into the rural South. When John

Williams Walker made inquiries to hire a tutor "who can read & write & above all speak French," he anticipated that the chief obstacle would be finding one willing to immigrate to Huntsville, Alabama. "*What* would induce such a one to come out here with us?" he wondered. Nothing, it turned out; Walker, like Calvert, eventually placed his daughter in a French-speaking school in Philadelphia.[40]

Critics of boarding schools objected to French's prominence in the curriculum because of its impracticality in English-speaking America and because of its aristocratic associations. Benjamin Rush argued that "the English language certainly contains many more books of real utility and useful information that can be read without neglecting the other duties by the daughter and wife of an American citizen." For a brief period in between the American Revolution and the French Terror, many Americans had developed a respect for French culture. Some figures even believed the young republic might develop closer relations with France than with Britain. After the Revolution devolved into the Terror, however, old prejudices against France resurfaced. During much of the antebellum period, with the exceptions of the revolutionary years of 1830 and 1848, France seemed to be the epitome of unsound politics, dissolute culture, and social instability. As one writer in a women's magazine charged in 1852, the taste of the French "is about as wretched as their morals." The language's popularity in boarding schools represented for most observers Americans' continuing fascination with aristocracy and European culture. "The study of the French language is, in most cases, a mere mania of the day," another critic argued in 1853. It contributed nothing to "making the home the centre and natural theatre of [women's] best graces" but was purely an aristocratic affectation, "since it finishes with the school days and never had any intended use" in any practical sphere of women's activity.[41]

All boarding schools—even middle-class institutions—offered courses in the language because French was the cultural and diplomatic lingua franca of the Atlantic world. After summarizing objections to French language courses, the *Putnam's* writer asked, "Would we then discourage the study? Far from it; we would only continue it through life; we would never undertake it without meaning to do so." No one disputed that studying French was necessary for travelers and for those interested in its literature and history. "I am convinced that no man can properly understand a people without knowing something of their language," wrote an American abroad in 1850. "Moreover it is an indispensable condition to comfortable travel." Critics objected to superficial exposure to the language, but they also condemned the immersion in French that conservative boarding schools offered. Its purpose was not practicality but the idealization of what respectable women and men saw as a corrupt, dissolute, and unrepublican society. Some among the gentry did, in fact, idealize the culture of the ancien régime and sought to transplant it onto American soil. After several years boarding at Sigoigne's academy, Maria Walker was ready to return to her plantation outside Huntsville. She boasted to her mother, "Mrs. Sigoigne says that I am a perfect French girl." She now preferred "the openness and urbanity of the French character" to Anglo-American moderation, "a reserve which excites my contempt not my respect," she announced primly. But young Maria had concerns. She doubted whether "backwoods" people possessed the sophistication to appreciate a manifestation of prerevolutionary France in their midst. "I expect all the people will stare," she concluded laconically.[42]

The very structure of the boarding-school experience was designed to produce "perfect French girls." Young women lived together for a year or more alongside French teachers and their families. Thus, they absorbed not merely the language but French morals and manners as well. They heard French music, became familiar with its fashions and literature, and learned about its history. Teachers encouraged their charges to write at least part of their letters home in French, both as practice and to display to their parents their progress in its study.[43] Many schools even discouraged or forbade their boarders from speaking English during the day. At Jennie Ellis's school, students spoke French "when addressing the teachers themselves or when at table." Isabella Mease assured her North Carolina friends that "the French language is learned so as to be spoken with ease & fluency in six months" at the Picots' academy. Perhaps the most genuinely "French" school of all was run by Anne Marie Sigoigne and her daughter Adéle. Students there not only attended classes in the language but lived a sort of colonial existence. As a rule, boarders "read and write French every day" and took "a lesson twice a week from a gentleman," according to Susan Gaston. "Of course," she added, "it is spoken in the house." Class work only passed on the mechanics of language. Appreciation for the reactionary principles behind the words and grammar required that young ladies become Frenchwomen, that their Walnut Street town homes become Parisian salons. Women not only had to speak French fluently but to live it—in class, over meals, in conversation, and in parties and salons.[44]

". . . the world and its customs"

French schools did not prepare their charges for lives as patriotic wives, mothers, and reformers of society. They educated privileged young women to embody the values of the European leisure class in American public spaces, such as concert halls, ballrooms, and salons. These aims confused and enraged middle-class education reformers. Almira Phelps charged that the *"fine ladies"* boarding schools "send into the world feel themselves ridiculously exalted above all sensible conversation, or all attempts to be useful." But French-school administrators and their patrons did not define utility in a way Almira Phelps would have recognized. Most institutions for women expected that their graduates would engage with the world. But they believed that women should do in a manner consistent with conventional understandings of femininity, which they assumed to be biological, divinely ordained, and immutable. Privileged folk also believed that gender was a physiological fact, not a social construction. But their definition of feminine propriety was anything but conventional. It flaunted middle-class nationalism by its idealization of Europe. Educational reformers objected so vociferously to French schools' social regimen not because it molded women to occupy public spaces, but because of the reactionary, even anti-American values they promulgated there.[45]

Planters examined a school's social contacts closely before they committed their girls' to their care. Classroom learning was important, but women would not pass their lives sitting at desks. They would need to shine at parties, the theater, and at other social occasions. Social life inside and outside the school was so pervasive that it constituted an informal curriculum where classroom skills could be honed and new ones learned. French schools boasted of their busy social calendar and their contacts in high society outside the school to attract patrons. In 1839 James Louis Petigru sent his daughter Susan from a Charleston academy to one in Philadelphia because he believed she "sees better society than she would do at home." In Philadelphia his daughter met frequently with the family of William Drayton, who, like Petigru—owing to their opposition to nullification—had become alienated from Charlestonians. In the Draytons, Petirgu enthused, Susan "could not have a better model nor visit a house by which she will improve so much."[46] Students often displayed far more interest in social activities than in the mundane rounds of class work. Y. E. Transou explained her failure to write her cousin about her life at a Philadelphia academy because, she claimed, she had no "interesting matter" to relate, although she went on for four densely packed pages about her observations of the belles and beaux at a nearby museum.[47]

The same local contacts who eased southerners' minds about their daughters' well being served as their guides into Philadelphia's social world. Their presence made the city's schools especially attractive to planter-class parents. Similar institutions could be found in more accessible southern towns, but the depth of social contacts that Philadelphia offered was not as readily available. Isabella Mease vouched for the Picots' school to her Iredell friends from North Carolina. The Iredells valued her friendship highly because her social rank gave her access to some of the city's most prominent families. As Alice Izard reported, Mease "has a very general acquaintance and seems to enjoy the good will of everybody." Because southern parents regarded exposure to the social world as an essential part of their daughters' educations, they were not reluctant to exploit their friends' "good will" for their benefit of their children. "In sending you to Philad[elphia]," William Polk informed his daughter Mary in 1824, he "had two objects in mind: the first for the improvement of your mind & acquirement of knowledge generally; the second, the attainment of Musick, drawing, French, and mixing with the fashionable world." He met his first goal by enrolling Mary at the Mallons' school. The second he advanced by using his letters of introduction to expose her to "the most respectable citizens of this place" and entreat them to invite Mary to their social affairs. Before leaving the city he boasted of his success to his wife, claiming that the "*fine, fine women*" to whom he had introduced Mary, including Mary Elizabeth Cheves, had been "pleased to say that they will send often for her."[48]

Not all parents looked forward to their daughters' entrée into polite society. The fashionable world was full of peril. Even upper-class parents feared that children might become entranced by its superficialities. As countless novels and short stories warned, the hothouse intensity of the beau monde threatened to turn modest young women into pleasure-loving coquettes.[49] William Gaston feared that this would be the fate of his daughter Susan. Joseph Hopkinson, under whose protection Gaston had placed her while at Sigoigne's academy, asked that Susan be allowed to accompany his family to social functions. Gaston demurred, fearing the influence of "dissipation" on his impressionable daughter. Susan, he wrote his friend, was already "too apt to fancy herself a woman." The heartless world of the parlor and ballroom would only accelerate this process. "My dear daughter has years yet in which I wish her to be considered, and to consider herself, a girl," he explained. Hopkinson empathized, but asked his friend to reconsider. Girls of Susan's stature would be public figures as adults—mistresses of a household, wives of men of affairs, and hostesses. They should not be sheltered from the world, but taught to master it. Susan, Hopkinson argued, ought "to be drawn into the world and its customs, among which, she is hereafter to live, than to be restrained from them." Good jurist that he was, Gaston knew when he was bested in debate. "I am well aware that many advantages may be derived from [social life] which she needs," he admitted, while restating his concerns about "excesses which an association with the fashionable and the gay might threaten." His reservations notwithstanding, Gaston relented. Given the social position to which Susan was entitled, he had no other choice.[50]

Boarding schools scheduled a variety of public activities so that young women could practice the skills they had learned. These festivities were part of the informal curriculum, since teachers monitored their charges' behavior. Most schools encouraged or required students to take an afternoon walk between the end of classes and tea. In smaller boarding schools, the entire student body participated en masse; in larger institutions, they embarked in groups broken down by age or class. In Philadelphia the route of choice was down Chestnut Street, the city's genteel shopping district. Here young women could observe the mannerisms of other refined promenaders and even compare themselves to the pupils of competing French schools. In 1817 students at Miss Lyman's school "consider[ed it] as a great deprivation" when their headmistress "prohibited the girls from walking" down Chestnut Street, "the most fashionable street in town." Teachers also escorted their pupils to parks, amusements, and neighborhoods where encounters with the lower orders could be contained. It was important for young women to learn how to handle themselves on the streets, where they were subject to the leers, curses, and insults of working-class pedestrians. Daily walks taught gentlewomen how to cope with this fact of Jacksonian American life. As the fictional Madame in Susan Petigru King's *Lily* explains to one of her students before an afternoon promenade, "Promenez-toi en demoiselle bien élevée, et ne regardez point les flaneurs."[51]

These excursions were not merely respites from the grind of the boarding-school regimen. Refined people occupied refined spaces; one's very presence there, "at the right moment in the right dress," argued Richard Bushman, "identified a person as genteel." So, in 1837 young Susan Polk of Murray County, Tennessee, reported with pride of her trip to Camden, New Jersey, with her classmates from the Miss Smiths' school, where they visited a "bewtiful grove" housing a "curcular railrode" that each student rode twice. Afterward, the young women retired to "the garden and got some Ice cream." The point of this trip was not just to provide the girls with amusement. Diversions, as well as their environments, had to conform to standards of refinement that distinguished those who enjoyed them from their social inferiors.[52]

Some students browsed or shopped during these excursions, although shopping was also a separate activity that engaged their time on weekends. It too had an educational purpose, however: to make women into knowledgeable consumers of fashionable goods. Emma Shannon and her fellow boarders from St. Mary's Academy "went tearing down Chestnut Street" one spring day in 1858 and "saw a great many sights," she boasted to her mother. But this was not a pleasure trip, as the girls were closely monitored by their teachers. They "are not willing to take us to any but the finest stores," she reported. Some of these excursions were minor, as when Helen Smith took her pupil Susan Polk shopping to buy a bonnet and some "colored dresses" on the termination of a period of mourning. But some French-school students spent money extravagantly on the clothes and accessories necessary to shine in a cultivated company. Lizzie Spragins's father outfitted her with nearly $200 worth of fashionable attire when he placed her at Sigoigne's academy in 1839, with license to spend more at her discretion. In 1841–42 the young Virginian spent almost $92 at five different shops on dresses, gloves, capes, and wraps. Parents sometimes balked at these expenses. When Sidney Gill sent Mary Wiley's father a bill for her spring 1854 wardrobe, he anticipated "that the total amount is larger than you have supposed it would be." Gill assured the Georgia doctor that Mary "seemed to require a good many clothes, and we know that you would wish her to have what was suitable and well-made." Because sociability was an integral part of French schools' curriculum, and because appropriate attire was absolutely essential to participation in polite society, such expenses were not indulgences. They were as basic as costs for books, board, and tuition.[53]

Some proprietors of French schools actually did possess high rank in prerevolutionary France or St. Domingue, so they had connections with prominent Philadelphia families. They used these links to lure wealthy students to their schools, and of course they had to exploit them if they wished to keep women enrolled from year to year. Teachers encouraged their pupils to write home often partly so that parents would know that their daughters were circulating in the city's parlors and drawing rooms. In 1823 Deborah Grelaud took her summer boarders on a day trip to the Woodlands, the estate of the Hamilton family on the banks of the Schuylkill River. It was "the most delightful place I ever was in my life," Georgia Bryan gushed to her grandfather. The elegant furniture, manicured grounds, and refined company described by Bryan announced to her family that she was not merely learning the genteel arts in school, but in practice. The Picots also enjoyed significant connections to Philadelphia families. A year after Annie Iredell left their school, Charles Picot filled a letter with the doings of the prominent

Pennsylvanians with whom she had grown intimate during her time at the school. The Picots also held affairs of their own, particularly concerts, to which they invited local luminaries. In this way students who did not enjoy preexisting connections with local families could be exposed to high society.[54]

Most French schools followed the Picots' lead in this regard. They held frequent receptions of their own because they could not be certain that all their students enjoyed access to Philadelphia families. Deborah Grelaud threw parties that featured visiting European musicians and vocalists, refined conversation, and a mixed company of students, visiting luminaries, and prominent locals. In 1840 Mary Middleton, the younger cousin of Eliza Middleton Fisher, attended school at Grelaud's and was looked after by her Philadelphia relation. Fisher had a standing invitation to attend Grelaud's musical parties, where she enjoyed the company of her young cousin and the skilled musicians who could always be found in Grelaud's parlor.[55] The Sigoignes, who had close connections to the Biddles, Fishers, and other well-placed Philadelphia families, held concerts and receptions where their boarders circulated with the city's upper crust.[56] It was at one of these receptions that Mary Jane Ellis of Mississippi caught the eye of Dr. René LaRoche, a virtuoso at the violin and piano who played for the Sigoignes' company. Ellis and LaRoche married in 1824.[57]

Few young women entered Philadelphia society expecting to find husbands. Rather, they wished to enjoy their time in the city and at the same time to prepare themselves to excel in company once they returned home. By exposing students to sophisticated Philadelphia women, teachers and parents sought to teach refinement by example. They also wished to demonstrate the benefits of genteel rank. Exposure to Philadelphia parlors and salons also served to demonstrate to southern girls the benefits of participation in this world. When Mary Wiley entered school in 1855, her cousin, who had attended the same institution a few years before, assured the nervous young woman that she too had feared "entering the world, of being hardened and chilled by its cold influences, of being infected by the corrupt atmosphere." Women of their station had a responsibility to overcome their natural reserve, she admonished her cousin. On her return she would assume her position as an "important member of the family circle, as a 'young lady' is necessarily obliged to be." But there was pleasure in privilege as well as duty. "It is pleasant to mingle in the world, and know that you are no longer a mere spectator of the busy scene but one of its actors." Novels and advice books stressed the artifice and competition of social life, but this writer stressed the class cohesion that elite education and the rituals of sociability produced. Fashionable life showed young women that they had "interests in common with the thousand bright eyes and brilliant forms around you, and that in spite of the declarations of wiser persons under the gay and apparently careless exterior of the fashionable, there are many hearts beating light with love and affection." Wealth and social position brought pleasures that gave the lie to the Cassandras who warned of the snares of polite society. They were not to be apologized for, but to be enjoyed.[58]

The rituals of polite society forged cultivated women and men into a leisure class. Inside and outside the classroom, French schools taught young women to assume the roles to which their wealth and social position entitled them. Parents and teachers wished, as one father stressed to his daughter, "to fit you to move with grace and credit to yourself in your appropriate sphere as a Lady." Critics charged that these lessons had no place in a society committed to egalitarianism, and they were right. But the families who patronized French schools did not subscribe to those ideals. If their fellow Americans could not be persuaded to embrace elitism, at least privileged people could follow these practices in their private companies. To do so cultivated women would need the company of equally refined men.[59]

4

The Republic of Medicine

If women en route to school in Philadelphia averted their eyes for a moment from the northern landscape to study the faces of their fellow travelers, they might have noticed some young men, equally nervous, in the seats and aisles around them. The Old South helped fill the classrooms of women's academies in Philadelphia, but it also sent its young men northward. Some of these boys, such as Tristim Lowther Skinner of Edenton, North Carolina, attended grammar schools in the city. These schools combined academic rigor with exposure to the social world. Far fewer young men than women attended socially focused institutions, however. The educational needs of southern boys were fundamentally different from those of their sisters. Women needed skills that would enable them to figure prominently in society, attract a desirable husband, and manage a household. Men also needed cultivation, but in addition they required attainments that would enable them to provide for a family and to participate in public affairs. Thus, southern boys usually learned at home the basic academic skills their sisters traveled to Philadelphia to acquire. They might head north, however, to pursue more advanced study. Southerners could be found amid the student body of every northern college and university of note; Princeton had a particularly large southern contingent.[1] Socially, southern men heading northward sought some of the same goals as the young women seated around them. Their academic and professional prospects could hardly have been more different.

Between 1800 and 1861 young men from every part of the Old South left home to study medicine in Philadelphia. The vast majority attended either the University of Pennsylvania (founded in 1765) or Jefferson Medical College (1824). In nearly every year they operated in the decades before the Civil War, a large minority or small majority of their students called the South home.

The record of these schools in drawing southern students is all the more impressive considering that it persisted in the face of two powerful countervailing trends. The first was an attack on the legitimacy of the medical profession and the educational institutions that maintained it. Most powerful during the early national period but extending late into the nineteenth century in practices such as hydropathy and homeopathy, this movement formed part of a popular revulsion against supposedly elitist vestiges of colonial life, such as lawyers and the educated clergy.[2] The second trend discouraging northern training was the partially successful effort of some southern physicians, politicians, and agitators to promote a "southern education for southrons."[3]

The young southerners who opted to study in Philadelphia could hardly have been unaware of these arguments. That they chose to ignore them suggests that, in their judgment, the merits of a Philadelphia education overcame the force of these objections—just as young southern women dismissed critics who condemned French schools as elitist and un-American. Three factors drew southern doctors-to-be to Philadelphia: the quality of its medical schools relative to their rivals in the slave states; the city's prosouthern atmosphere; and the opportunity to acquire social status along with a medical degree from a prestigious institution. The high quality of the University of Pennsylvania and Jefferson Medical College compared with their rivals within the United States was universally recognized. The notion of a regionally based medical education appealed to southerners, but until the late antebellum years few schools in the South could rival the quality of instruction and practice available in Philadelphia.

The other reasons for the popularity of Philadelphia medical schools among southern men were obscured by the fog of sectional politics and egalitarian rhetoric, and they have remained underappreciated. The prosouthern atmosphere of the city little resembled the dire depictions of the North drawn by regional partisans. Sectional agitators warned that young minds would be corrupted by exposure to northern "isms" (abolitionism in particular). Even though advocates for southern medical education for southern youth lumped Philadelphia alongside Boston and other centers of radicalism, physicians-to-be and their professional mentors knew better. Philadelphia was understood to be a safe destination for the young men of the South. In addition elite southerners did not conceive of medical practice strictly in professional terms. They expected their sons to master social skills and cultivate contacts as much as, or more than, they wished them to achieve technical competence. In short a Philadelphia education promised to turn southern men into American gentlemen, not merely physicians. For the most ambitious southern students,

Philadelphia's major schools promised entrée into a national circle of gentlemen united by cultivation, social status, and professional interest—a community that Benjamin Rush, the dean of Philadelphia doctors, dubbed the "republic of medicine."[4]

"The atmosphere of Philadelphia is medical"

Southern men who wished to become professional physicians—holders of the M.D. degree—were not compelled to study at Philadelphia. True, until the mid-1820s the only medical school of any size in the South was the Transylvania Medical College in Lexington, Kentucky. After that, however, medical colleges and medical journals, such as the *Georgia Blister and Critic* (1854–55) and *Southern Medical and Surgical Journal* (published in three series from 1836 to 1867), began to be founded across the region. The New Orleans College of Medicine, the Louisville Medical Institute, and colleges in Georgia, Tennessee, and South Carolina all matriculated significant numbers of men by the 1840s. Yet many southern schools faltered because of low admissions, and not a few failed.[5] These institutions confronted Philadelphia schools with increased competition (which also came from new medical schools in the North and West) and frustrated efforts to impose rigorous standards on medical practice, such as European-style curriculums, admissions standards, and state-mandated licensing requirements.[6] Many of these schools modeled their curricula and clinical instruction on those of the Philadelphia schools—which was not surprising since a significant proportion of their faculty had been educated there. Southerners also had the option of studying in Europe. Only the most ambitious and wealthy could do so, but their numbers were not insignificant. The destination of choice during the early nineteenth century was Paris, whose innovative methods pushed Edinburgh, the eighteenth-century favorite, into the background. At least seven hundred Americans, about half of them southerners, attended lectures in Paris from 1820 to 1861. After the Civil War the center shifted eastward to Germany.[7]

Despite the proliferation of respectable alternatives, Philadelphia schools retained their standing among southerners considering a conventional medical education. In fact, the University of Pennsylvania and Jefferson Medical College became increasingly popular among southern men throughout this period. Table 1, which details the graduates of the Jefferson Medical College from its founding through 1850, shows that the school grew steadily more popular with southern doctors. Just under 140 men took degrees in its first five classes. After just ten years of operation, however, it was graduating more

Table 1: Graduates of Jefferson Medical College, 1826–1849

	1826–30 *N=139*	*1831–35* *N=196*	*1836–40* *N=505*	*1841–45* *N=385*	*1846–50* *N=928*
Penn.	57.55%	41.83%	28.91%	34.80%	27.16%
Foreign	3.60	4.08	1.38	3.12	1.83
North*	27.34	27.55	25.94	21.82	16.49
Del.	3.59	1.53	1.78	1.30	1.08
Md.	1.44	2.55	6.34	2.34	2.91
Va.	2.88	11.22	17.82	18.70	26.29
N.C.	.72	1.53	2.77	2.86	5.17
Ky.	1.44	2.04	1.98	2.34	2.16
Tenn.	0.0	1.02	3.17	1.04	3.45
S.C.	.72	2.04	2.79	2.60	1.72
Ga.	0.0	3.59	3.56	3.90	4.53
Miss.	.72	0.0	1.19	2.60	2.05
Ala.	0.0	.51	1.19	2.06	2.59
La.	0.0	0.0	.79	.26	.43
Fl.	0.0	0.0	0.0	.26	.54
Mo.	0.0	.51	.39	0.0	1.51
Ark.	0.0	0.0	0.0	0.0	0.09
Southern	11.51%	26.54%	43.77%	40.26%	54.52%

*N.J., N.Y., Conn., Mass., Maine, N.H., Vt., Ohio, Mich., Wisc., Iowa, Ind., D.C., U.S. Army & Navy.

Source: Frederick B. Wagner Jr. and J. Woodrow Savacool, *Thomas Jefferson University: A Chronological History and Alumni Directory, Annotated and Illustrated, 1824–1990* (Philadelphia, 1992).

than one hundred doctors every year. Only 11.5 of these men were southerners during the years 1826 through 1830, but in the next five years the percentage more than doubled, to 26.5 percent. By 1846–1850 more than half the graduating class—more than 500 men—were southerners. Virginians led all southern states. Their numbers gradually increased until, by the 1846–1850 period, they represented more than 26 percent of all graduates. Their numbers rivaled even the native-state contingent. By 1847 Virginians outnumbered Pennsylvanians 46 to 44. Philadelphia was readily accessible to men from the upper South, and this proximity was attractive since the only well-established southern schools in this period were in Louisiana and South Carolina. After 1840 Virginians consistently approached the numbers from all the northern states combined, less Pennsylvania. In 1846 through 1850, when about a quarter of the graduating class were Virginians, only 16.5 percent came from northern areas other than Pennsylvania.

As sectional tensions mounted in the 1850s, southern medical journals admonished young men not to humiliate their region by seeking an education outside it. Yet more southern men than ever took seats in Philadelphia lecture halls during these years. Southerners occupied more than half the desks during the 1850s, a figure that fell to 48 percent in 1860–61—an impressive figure coming a year after the highly publicized exodus of southern students from northern medical schools. Table 2 examines the last antebellum decade in more detail, breaking down the entire Jefferson student body by state. In 1859–60 nearly 70 percent of the class was southern. More Virginians were enrolled than any other state except Pennsylvania. More Georgians than ever attended classes, coming to 10 percent of the student body. North Carolinians too came in ever-larger numbers, never dipping lower than 5 percent of the class and composing 10 percent of the 1861 class. From the Deep South, Mississippi and Alabama were well represented, together claiming 16 percent of the seats in Jefferson lecture halls in 1859–60 and 9 percent four years before. Farther north, Kentucky and Tennessee posed comparable numbers. Their low point was the 1855–56 session, when they represented approximately 2 and 3 percent respectively of the Jefferson class, though their 25 students totaled one more than the combined numbers of students from New York and New Jersey. Professors could expect to instruct a few students from Florida, Louisiana, Texas, Arkansas, and Missouri every year.

Tables 3 and 4 show that the University of Pennsylvania was at least as successful as its younger counterpart in attracting southern men. Its 1859–61 sessions, when a healthy 39 percent of its student body was southern, represented the antebellum low point. From 1831 to 1840, more than 3 of every 5 Penn students had come from the South to learn their craft. During its early years, the Jefferson Medical College catered mainly to Philadelphians; only 11 percent of its graduates were southerners between 1826 and 1830. At this time more than half of University of Pennsylvania medical students were from the slave states. Of the more than 1,300 students in the 1850–53 classes, more than 51 percent were southern. While the University of Pennsylvania seems to have attracted more men from North Carolina, particularly at midcentury, the state-by-state breakdown closely mirrors Jefferson's patterns from the mid-1840s

Table 2: Matriculation, Jefferson Medical College, 1850–1861

	1850–51 *N=504*	*1855–56* *N=510*	*1859–60* *N=630*	*1860–61* *N=443*
Penn.	31.15	27.25%	19.21%	31.60%
Foreign	1.39	1.96	.95	2.03
North*	15.87	16.27	11.43	18.74
Del.	1.98	1.12	1.27	1.13
Md.	.99	.98	2.38	3.16
Va.	18.45	11.96	14.92	8.35
N.C.	6.36	7.89	6.98	9.71
Ky.	4.17	1.61	3.49	4.74
Tenn.	2.78	3.33	5.55	3.61
S.C.	2.58	4.31	5.71	.68
Ga.	5.16	10.39	6.98	4.74
Miss.	2.18	2.74	7.78	2.90
Ala.	3.37	5.88	7.93	2.26
La.	1.19	.39	.32	0.0
Fl.	0.0	.20	.63	.22
Mo.	1.39	1.37	1.90	3.20
Tex.	.79	1.37	1.14	1.58
Ark.	.20	.98	1.43	1.35
Southern	51.59%	54.52%	68.41%	47.63%

*N.J., N.Y., Conn., Mass., Maine, N.H., Vt., Ohio, Mich., Wisc., Iowa, Ind., D.C., U.S. Army & Navy.

Source: *Catalogue of the Trustees, Professors, and Students of the Jefferson Medical College of Philadelphia* (Philadelphia, 1851–61).

through 1861. It is important to observe that during this same period southern schools attracted a negligible number of northern students. In the 1830s and 1840s, when sectional antagonism was at a low pitch, Philadelphia schools matriculated a respectable body of southern men. They improved on this record as sectional tensions mounted and as southern doctors honed their arguments for regional medical education. In 1860–61, on the eve of the Civil War, Jefferson and Penn, the two largest schools in Philadelphia, matriculated 908 students. More than 40 percent of them were southern.

Table 3: Graduates of the University of Pennsylvania Medical School, 1806–1850

	1802–10	*1811–19*	*1826–27; 1828–30*	*1831–40*	*1841–50*
	N=288	*N=564*	*N=354*	*N=1449*	*N=1593*
Penn.	20.83%	19.68%	31.36%	20.01%	21.60%
Foreign	2.77	1.78	.85	2.00	1.76
North*	11.80	8.86	12.99	11.11	11.17
Del.	2.77	2.49	3.67	1.79	1.44
Md.	12.84	7.27	2.54	2.42	3.33
Va.	28.12	39.00	31.07	30.57	24.36
N.C.	1.39	3.55	5.93	9.25	11.42
Ky.	3.13	3.90	.58	1.38	1.13
Tenn.	1.39	.35	.28	3.52	5.46
S.C.	11.15	8.15	5.65	4.49	3.26
Ga.	3.47	3.72	4.80	5.38	2.95
Miss.	.34	1.06	.28	2.00	3.39
Ala.	0.0	0.0	0.0	3.80	5.15
La.	0.0	.19	0.0	1.86	1.07
Fl.	0.0	0.0	0.0	.42	.50
Mo.	0.0	0.0	0.0	0.0	1.76
Tex.	0.0	0.0	0.0	0.0	.19
Ark.	0.0	0.0	0.0	0.0	.06
Southern	64.60%	69.68%	54.80%	66.88%	65.47%

*N.J., N.Y., Conn., Mass., Maine, N.H., Vt., Ohio, Mich., Wisc., Iowa, Ind., D.C., U.S. Army & Navy.

Sources: "Recapitulation of the Number of Graduates since the Year 1802 [through 1819];" "List of Medical Graduates in the University of Pennsylvania from the Year 1814 to 1850, ca. 1814–ca. 1850," Philadelphia File Folder 1574, Box 23, UPA 3: General Administration (University Archives and Records Center, University of Pennsylvania, Philadelphia).

As these patterns suggest, many southern men found the idea of an urbane medical "republic" more alluring than the politicized version propounded by sectional agitators. Their arguments probably persuaded some young men to attend southern institutions. It is clear that many prospective doctors remained unconvinced, however. A combination of factors influenced young men as

Table 4: Matriculation, University of Pennsylvania Medical School, 1850–1861

	1850–53	*1853–56*	*1856–59*	*1859–61*
	N=1306	*N=1260*	*N=1297*	*N=993*
Penn.	36.83%	32.38%	32.54%	45.32%
Foreign	1.84	2.46	4.47	3.73
North*	10.18	10.00	12.10	12.08
Del.	2.45	1.90	1.69	1.81
Md.	2.22	1.50	2.00	3.52
Va.	12.25	11.03	8.40	7.55
N.C.	9.19	14.04	15.34	11.60
Ky.	1.84	.95	1.77	.81
Tenn.	6.35	5.75	5.01	3.32
S.C.	1.45	2.78	3.08	1.81
Ga.	3.67	2.14	1.85	.90
Miss.	4.98	5.48	3.32	2.42
Ala.	4.13	5.40	6.71	2.82
La.	1.22	1.27	.31	.70
Fl.	.25	.22	.56	.10
Mo.	.62	1.11	.31	.81
Tex.	.15	.48	.31	.30
Ark.	.38	1.11	.23	.40
Southern	51.15%	55.16%	50.89%	38.87%

*N.J., N.Y., Conn., Mass., Maine, N.H., Vt., Ohio, Mich., Wisc., Iowa, Ind., D.C., U.S. Army & Navy.

Source: *Catalogue of the Trustees, Officers, and Students of the University of Pennsylvania* (Philadelphia, 1850–1861).

they considered which institution to attend, but there can be little doubt that one factor stood out: professional physicians across the United States rated Philadelphia's two schools as the nation's best. Both boasted outstanding faculties, attractive classrooms and buildings, and ample opportunities for clinical instruction and hands-on experience.

Southern medical reformers confronted these advantages with a number of arguments. Some slave-state physicians maintained that the region's distinctive disease environment, its rural character, and its African American labor force rendered lessons learned in Philadelphia irrelevant to everyday medical practice in the region. The faculty of Penn and Jefferson were instrumental in fashioning the doctrine that diseases and their treatments were highly contingent on local environmental factors. Thus, they unintentionally gave succor to the movement for regional medical education. "There is a distinction between Northern and Southern medicine," insisted a New Orleans doctor, "as surely as there is a distinction between foreign and American medicine." Men who planned to practice in the South should study there, since much they would learn in Philadelphia or elsewhere would be ineffective, or even harmful, when applied back home. The argument that doctors ought to train in the region in which they would practice had a solid foundation in antebellum clinical doctrine.[8]

However, critics of the Philadelphia schools did not base their arguments on medical grounds alone. There were also questions of southern honor and regional intellectual culture to consider. Some southern doctors sensed that students' preference for Philadelphia reflected badly on themselves. Making an appeal for the New Orleans School of Medicine, a southern writer charged that "the annual pilgrimage of Southern young men to the medical schools of the North is an unnatural and humiliating sight." Many men were simply unwilling to sacrifice their medical training or social prospects at the altar of southern pride, however. John Dabney, a Virginia man, was typical. He felt a profound bond to his native state, yet he resolved to stay in Philadelphia. "I must sacrifice my inclination to my interest," he declared, "as I shall have a much greater opportunity here for acquiring information than I could have in Virginia." More practical critics charged that Philadelphia schools drew vital resources out of the region, enriching the North at the South's expense. A Virginian writing in the *Southern Literary Messenger* in 1842 argued that "to those who are desirous of the advancement and improvement of the community," medical schools were "objects of deep interest, and they are not the less objects of concern, to those who regard them merely as a means of increasing its wealth." The nearly 600 students from throughout the South enrolled in Philadelphia medical schools that year spent $350,000, he estimated. As if the exodus of hundreds of young men was not indignity enough, the loss of their fees, boarding costs, and other expenses provided further evidence of the South's relative weakness in the Union.[9]

The southern attack on northern medical schools was intense. It reflected the drift of sectional relations in the years preceding the Civil War. Those facts have served to obscure the very different opinions held by leading figures in the southern medical establishment, whose views more accurately mirrored those of the region's doctors. These men, many of whom had been trained in Philadelphia, did not feel threatened by the city's preeminence—quite the contrary. They felt a strong bond of loyalty to the institutions where they had been initiated into the medical community. When Daniel Drake planned to establish a medical school in Cincinnati to serve the southwest, he thought purely in colonial terms. "I am convinced that Cincinnati is to be the *Philadelphia* of the West as to Medical Instruction," he told his friend Samuel Brown. "We must be contented to compose a school of preparation for our friends & masters the faculty of the Phila. school." Just as American doctors in the eighteenth century had finished their training at Edinburgh, Drake hoped that "a considerable number of young men after attending lectures here will visit Phila."[10]

Richard D. Arnold, a Savannah, Georgia, physician, expressed similar sentiments after his election to the College of Physicians of Philadelphia. "To Philadelphia our profession in other parts of the Union looks for the beacons to guide us onward," Arnold wrote in thanks. "A Fellowship with such lights of our profession as exist among you is then a matter of honorable satisfaction." An honest reckoning compelled southern doctors to recognize Philadelphia's superiority, but something more was at work. Sophisticated gentlemen viewed sectional ardor as evidence of small-mindedness. Advocates of "narrow provincialism"—sectional firebrands or breast-beating nationalists—were disdained as vulgarians. Seeing themselves first as American gentlemen, not southern doctors, socially prominent physicians regarded Philadelphia not as hostile territory but as the center of their professional community.[11]

Such loyalty owed much to fond memories, personal bonds, and cosmopolitan ideals, but those feelings would have amounted to little if Pennsylvania and Jefferson had not maintained their academic standing. Both schools enjoyed a reputation for academic rigor. The curriculum changed over the course of the early nineteenth century, but at the University of Pennsylvania, the nation's leading institution, students took two courses of lectures, the second a continuation of the first, and wrote a thesis if they wished to obtain a degree (many students matriculated and attended classes without graduating, however). Normally it took three years to obtain an M.D. degree. Students attended lectures in two four-month terms each year, studying seven courses, which at various times included anatomy, physiology and pathology, materia medica, chemistry, theory and practice of medicine, surgery, diseases of women and children, botany, natural and experimental philosophy, midwifery, and pharmacy.[12]

Pre–Civil War medical-college students were renowned for their disdain for academics, and historians have taken a dim view of the quality of antebellum medical schools, particularly when compared with contemporary European institutions. Yet many students approached their studies with the utmost seriousness. Young men became drawn into the routine of classes, surgical observations, and clinical work. "It is enough to kill a horse almost to set 8 hours study every day upon hard benches I feel very much like resting at night but there is no rest for the wicked, or righteous here all fare alike," complained Neil McNair, a North Carolina student, in 1838. His first year amounted to "one Anatomical-Surgical-Physio-Chemical sing song from morning to night." Hyperbole aside, the young man was not far off the mark. Students might attend lectures for up for eight hours a day. In addition evening reviews offered by faculty and senior students were popular among serious scholars. Robert Nelson of Jefferson County, Virginia, found himself and his roommates "engaged as long as ten hours per diem attending lectures and quizzes, and besides this we have the lectures to study so you may guess that our time is well occupied." Some southern medical schools provided a comparative quality of instruction, but William Penn's city was different. It offered an integrated medical, professional, and social experience that no southern school could match. "The atmosphere of Philadelphia is medical," observed Charles Bonner in 1842. Like legions of other southerners before and after him he was "determined to remain within its influence."[13]

Most professionals believed the city's greatest advantage over its southern rivals was the availability of hands-on instruction in clinics and hospitals. Philadelphia was by no means unique in making clinical work available to students. Urban medical schools, North and South, enjoyed this advantage over their rural counterparts. However, Philadelphia's great size relative to southern cities and towns rendered it far superior in this respect. "As for learning the theoretical part of medicine and studying those branches on which we much found our practice (as Anatomy & Physiology) Philadelphia presents as many advantages perhaps as any other place in the United States," a Virginia student asserted. Josiah Nott, an Alabama medical reformer, told a protégé that he would "incalculably benefit" from attending Jefferson Medical College because of the likelihood of being "enabled to attend the surgical wards." Regional advocates warned that treating northern ailments would do southern practitioners little good, but students paid them no heed. In justifying his choice of the University of Pennsylvania over a southern school, Charles Bonner summarized Philadelphia's advantages over its southern counterparts. The city, he explained, offered "the best talent of the country in its collection & in its anatomical & operations Library and chemical apparatus." Those men who pursued both medical skill and gentility could hardly overlook those considerations.[14]

" . . . warm friends of the South"

Philadelphia's popularity with southern pupils cannot be explained by the quality of its medical schools alone, however. Parents would not have sent their sons into an environment where they feared for their moral and physical safety. Yet the southern press was rife with just such dire warnings about conditions in the North. Some students traveled there in spite of this alarmism, resolved to insulate themselves as best as they could from abolitionist propaganda. Others, however, knew about Philadelphia's longstanding ties to the region and its conservative atmosphere. Whatever their presuppositions about the city, once there students discovered that Philadelphians were nearly as hostile to antislavery agitators as they were. Though tensions sometimes roiled relations between locals and students, these conflicts were almost always town-and-gown issues rather than sectional disputes.

True, some students attributed Philadelphians' hostility to antisouthern prejudice. Edward Warren, who was trained at Jefferson, recalled that he and his fellow scholars "were popularly rated and reviled as Southerners. . . . The result was a perpetual state of warfare between the Philadelphians and the [students]." But Warren's feelings, recorded years after the fact, are tainted by his bitter experience as a Confederate doctor. Clear-eyed southerners admitted that, more often than not, Philadelphians' aloofness stemmed from student rowdiness. As one perceptive young doctor observed, "young men in a city with nothing to do are not very apt to do nothing." Witnessing the daily spectacle of students bailed out of the city's watch house, he laid the blame squarely on his fellows' shoulders. "I am very sorry to say that I think nine times of ten the students are in fault." In fact, he praised Philadelphians for their forbearance. "I have been in a great many places in this city and never have I been the least insulted," he admitted. "At first I thought we were imposed on but now I think the contrary." Sore feelings spoiled relations between some locals and students, but not because of sectional biases on the part of Philadelphians. If southern men felt some resentment, a North Carolinian observed, it was because students "deem it their especial prerogative, to create as much mischief, and to play the devil as much as possible."[15]

Southern partisans believed that exposure to an antislavery culture threatened young men more than the alleged hostility of Philadelphians. James C.

Billingslea accused northern doctors of looking down upon him and other southern students "simply because we are *slaveholders.*" Harvey L. Byrd lamented the large southern presence in the "crowded lecture rooms" of Philadelphia, where impressionable young men fell under the sway of "*Northern men,* who for the most part are *inimical* to the South and her 'peculiar institutions.'" Better, charged J. D. B. De Bow, that southern men "remained in honest ignorance and at the plough-handle," than exposed to "doctrines subversive of their country's peace and honor." Southerners in Philadelphia quickly learned that these charges had very little foundation. The city did not teem with abolitionist sedition. In fact, southern men took heart at the obvious disdain with which Philadelphians viewed antislavery activists. As the Philadelphia-born journalist Charles Godfrey Leland recalled, "there was hardly a soul whom I knew . . . to whom an Abolitionist was not simply the same thing as a disgraceful, discreditable malefactor." Little wonder, then, that young men ignored the dire warnings of southern medical alarmists. The local population received southern students warmly, and when the issue of slavery did arise, Philadelphians affirmed the students' deeply held social and racial attitudes.[16]

Some conflict could not be avoided, however. No issue rankled students more than the well-publicized activities of the small band of local abolitionists. Students took offence at antislavery actions that insulted their honor as southern men. One pious Virginian raged that "any southern person to go to church here . . . may confidently expect about every other Sunday to have his feelings outraged on a subject that I think of all others should be kept out of the pulpit." He vowed to attend the Presbyterian church of Dr. Henry Boardman, a staunch Union man who "never alludes to the abolition question in the slightest degree," and to avoid the closer church of Dr. Albert Barnes, who "carries his abolition sentiments with him into the pulpit." Such apostasy would have been unthinkable in a southern church. Slave-owning men, accustomed to hearing the peculiar institution extolled as blessed in the eyes of God, might well have been shocked.[17]

Yet the radical clergy posed only slight difficulties. After all, they outraged only churchgoing young men—a minority of the student population—and the godly could always choose a more congenial church. Everyday encounters with antislavery activists posed a more intractable problem. "It is not a very unusual sight to see a beautiful young Quaker lady escorted in the streets by some of the colored beaux," David Hamilton exclaimed to his cousin at home in Athens, Georgia. Southern men, accustomed to the rigid, deferential social relations enforced by slavery, adjusted with difficulty to the North's less formal codes of interracial behavior. "One of the Miss Grimkys was married not long since," he continued, gathering steam, "& at the party the whites & blacks mingled promiscuously." For the Georgian, as for other planter men, it was all a bit hard to bear. More than any other aspect of northern life, the presence of abolitionists made Philadelphia seem like foreign territory. "I am a southron in soul & feeling," he averred, "& when I again [visit] my native soil, there will be gladness of heart & abundant rejoicings."[18]

Despite their disdain for the North's more liberal racial mores, most southern students soon found their misgivings assuaged. Scenes of racial egalitarianism offended them, but they were infrequent, to say the least. Moreover, Philadelphians showed their true feelings and soothed southern tempers by ignoring, insulting, and sometimes assaulting abolitionists. Hamilton rejoiced when a crowd of locals and southerners razed Pennsylvania Hall, the new antislavery meeting place, in 1838. "Our city," he exulted, "has again resumed its characteristic order & calm & quiet after the tumultuous outbreaking of an enraged populace. An all wise & just God has kindly made mobocracy the instrument of his wrathful visitation." As Sidney George Fisher had also observed, the crowd shouted down the speakers, beat up sympathetic spectators, and set fire to the building, all while the police looked on. "It was the grandest spectacle that I ever before witnessed," Hamilton testified. "Philadelphia has by the late proceeding raised herself in my esteem, although she before held a high station in my affections." Twenty years later another student reported that local animosity toward antislavery radicals remained intense. "The abolitionists hold meetings every few days," he observed, noting the attendance of southern students and "a large number of citizens," whose hisses, coughs, and catcalls drowned out the speakers. As in 1838, the student colony enjoyed the collusion of the authorities in enforcing racial subordination. "I tell you they make the free negroes walk a straight line," this Carolinian boasted. "One of the students knocked one down the other day and beat him like the notion and the police stood and never said a word."[19]

Extreme manifestations such as these of Philadelphians' antipathy toward the antislavery movement clearly heartened southerners. The prosouthern attitudes and conservative atmosphere that students encountered daily put them at ease far more effectively, however. Carrie Fries visited Philadelphia in 1860 to attend her fiancé's graduation from Jefferson. Shocked to learn that the polite young woman with whom she was conversing at her hotel was an abolitionist, she "got a little excited" until she discerned that the rest of the company viewed the woman as a harmless fanatic. "All the other persons that I have seen here are warm friends of the South," she reported to her family back home. Philadelphians felt a kinship with the South that they did not extend to other regions, particularly New England. These feelings apparently led Philadelphians to establish a special bond to southern students, whom they simply labeled *"Virginia Doctors."* Harriet and Georgina Manigault reported "that those beings are very much prized here."[20]

Besides the congenial local population, the large cohort of southerners also helped students from the slave-owning regions feel less isolated. If they wished, young men could find solace in the company of friends from the slave states. One student described Professor Nathaniel Chapman of the University of Pennsylvania as "a real Virginian . . . [who] has not deserted old Virginia principles." Not only was the elderly doctor "very polite indeed" to southerners, but upon hearing that a group of Virginians had been hauled into court after a row, Chapman "went immediately to the Mayor's office and released them by giving security to the amount of ten thousand dollars." Little surprise, then, that the shrill cries of medical sectionalists rang hollow in Philadelphia. Spending a year or two there was not so different from living in the South, after all.[21]

It is in this context that the exodus of 244 students from Penn and Jefferson in the wake of John Brown's execution for attempting to lead a slave revolt at Harper's Ferry, Virginia, in October 1859, must be understood. On the surface, the "secession" (as it was called) suggests that relations between the city and its southern guests had deteriorated by the late 1850s. The aftermath of the affair, however, revealed the incident to be more of a college prank than a prelude to the republic's dissolution. Ostensibly the students left to protest the tense atmosphere following Brown's hanging and the arrest of several armed students at a lecture by abolitionist George W. Curtis. On December 17, 1859, the Medical College of Virginia in Richmond received three telegrams from southern students in Philadelphia. The first two claimed that 150 young men were prepared to leave Philadelphia for Richmond if the Medical College there would admit them through the current term. As the college's board met to consider the proposal, a third telegram arrived, pleading "We anxiously await your reply. For God's sake let it be favorable—only diploma's fee. We are in earnest, confidential." The board voted unanimously to admit the students tuition free. At the Medical College of the State of South Carolina, Dean Henry R. Frost approved a similar request from Carolinians in Philadelphia.[22]

The students fled Philadelphia after receiving word from Virginia that they would be admitted. They arrived in Richmond to a "hero's welcome" led by

Virginia's fire-eating governor, Henry Wise. Of the 244 students who arrived in Richmond, all but a hundred enrolled at the Medical College of Virginia. The rest moved on to the Medical College of the State of South Carolina, other schools, or home. Their withdrawal delivered a serious blow to Philadelphia's schools. It hit Jefferson particularly hard; 69 percent of its 1859–60 class was southern. Only 40 students left the University of Pennsylvania.[23] After the stunt's initial shock had passed, however, it became clear that as many southern men remained in Philadelphia as had left. Their presence undermined the credibility of the seceding students and their southern coconspirators because they "served as visible reminders of the city's conservative character." Surely, observers noted, if Philadelphia had been such a hostile environment, few southerners would have remained. The press claimed that the students who stayed in Philadelphia represented "by far the most respectable" southern members of their classes. As the Unionist counterattack gathered steam, critics condemned the Richmond faculty's opportunistic collusion in the affair, Wise's gloating, and financial incentives to the students supplied by the Common Council of Richmond. That body had contributed more than thirty-five hundred dollars to cover train tickets and other expenses. Such generosity did nothing to advance southern medicine or sectional reconciliation but instead inflamed "the excited feelings of great portions of our country," moderate voices pointed out. A "Unionist" writing on the secession in the *Baltimore American* insisted that "no conservative man, by whom I mean a union-loving citizen as much opposed to the fire-eating Southerner as the fanatical Northern abolitionist . . . should countenance, by any act, that which would . . . turn the balance" toward disunion. If the incident exposed sectional tensions, it also reinforced the nationalist spirit represented by the large population of southern students in Philadelphia.[24]

The termination of the affair deflated even the modest hopes of southern radicals. Reports circulated that made the seceding students seem irresponsible and immature. The *Baltimore American* reprinted a column first published in a Philadelphia paper reporting on the return of a number of students, who admitted "that their visit to Richmond was only to enjoy a Christmas frolic, and an exhibition of Governor Wise's oratorical pyrotechnics." The "treason" of these backsliding students enraged the editor of the *Oglethorpe Medical and Surgical Journal,* Harvey L. Byrd, who marveled how young men "at the most critical period of their country's history, [could] patronize the institutions of their enemies." Charges of betrayal aside, these young men were not alone: southerners continued to choose Philadelphia medical schools instead of those within their region. The year after the heavily publicized secession, 34 percent of the University of Pennsylvania's class and 48 percent of Jefferson's were southern. Though both were significant declines from previous years—Jefferson's southern enrollment was 69 percent in 1859–60, the antebellum high point—it was a healthy figure given the publicity generated by the student secession. Philadelphia's conservatives had good cause to believe they had preserved the trust of the South.[25]

". . . the medical brethren"

Southerners had good reason to feel at home in Philadelphia, but the city's attractions extended well beyond its highly regarded schools and friendly milieu. Colleges, universities, and professional schools provided young men with the opportunity to acquire contacts, polish, and other qualities besides an education. But Virginians and Carolinians did not have to study in the North to become genteel. Many of the young men who attended Penn and Jefferson aspired to more than membership merely in a regional gentry. Study in Philadelphia had the potential to offer initiation into two exclusive communities: the society of medical gentlemen (Rush's "republic of medicine") and the even more exclusive world of the American gentry. Some southern men chose to attend schools in Philadelphia to enhance their reputations among the upper class and to secure membership in a nationwide community of professionals. Access to this group, in turn, promised entry into the world of the urban establishment.

Certainly, sophisticated slave owners were frustrated by the region's anti-intellectualism, cultural underdevelopment, and pervasive individualism. Such men longed for the excitement, cultural diversity, and urbanity that only urban areas—particularly the great cities of the Northeast—could provide. Moreover, as southerners and gentlefolk they craved the honor and affirmation conferred by the recognition of their peers. Southern "assumptions of status, taste, and good breeding . . . were quite compatible with the criteria prevailing in the salons of Boston, New York, Philadelphia, and London," so the opportunity to study and socialize in Philadelphia could hardly fail to excite socially astute young men. Of course, such men represented only a minority of southerners attending classes in Philadelphia. Most Americans looked askance at anything tainted with the faintest whiff of aristocracy. However, the choice to pursue a conventional medical education at a leading institution in an era of rampant sectarianism left students open to accusations of elitism. And while only a minority of students pursued gentility along with medical proficiency, they were an important group. Their horizons were national, not local or regional; they aspired not merely to practice medicine, but to lead their profession and engage with the wider world.[26]

The fact is that many observers who identified the city and its medical institutions with aristocracy were dazzled by the association. Gentlemen found the egalitarian critique of professional education socially repugnant as well as medically unsound. Parents and young men judged medical schools according to their social qualities, not merely their educational features. Students made the long journey to Philadelphia's lecture rooms to become gentlemen as well as doctors. In 1818 forty-four Penn students made these concerns manifest in opposing the candidacy of Robert Hare for the chair of chemistry. "Natives of distant sections of the Union," they had "repaired to Philadelphia . . . to aspire to the highest professional honours this country can bestow." These students feared that Hare's candidacy might damage the school's standing, thereby tarnishing their reputations. They argued that the well-connected Hare owed his appointment to his social position, not his professional credentials. "There is connected with every profession an honorable pride, without which it would be spiritless, groveling, and worthless," they reminded the trustees. As their opposition to Hare reveals, even socially adept students did not uncritically glorify high rank. If they did, Robert Hare would not have excited their opposition. Rather, students took their professional identity seriously. Their sense of themselves as American gentlemen was inseparable from their calling as doctors.[27]

As the students' concerns suggest, Philadelphia schools attracted rural young men hungry for professional as well as social status. A Kentuckian reported that both sophisticated southerners and Philadelphians mocked students from Tennessee and his state as "Back Woods boys" because of their country manners. Such young men flocked to northern medical schools from town and country alike, rich and poor. Some of these doctors-to-be merely sought the patina of respectability that study in Philadelphia conferred. Even if they did not stay to earn a degree, they hoped that their experiences would elevate them above the quacks, sectarians, and untrained practitioners who competed in the antebellum medical marketplace. Some of these young men hoped that a year or more in the city could turn them into sophisticated professionals with national connections. The socially ambitious went to Philadelphia because its schools were associated with elitism, not despite it.[28]

Established southern gentlemen also craved fellowship with the Philadelphia community because of its association with cosmopolitanism and privilege. When Richard D. Arnold of Savannah agreed to help his Philadelphia colleague Alfred Stillé obtain a position, he revealed the urbane sensibility that tied leisure-class southerners to their northern peers. He also betrayed the contempt that many sophisticated planters felt toward their home region. "As a mere provincial, a kind of outside professional Barbarian, I feel flattered that you should have thought I might in any way contribute to that end," Arnold wrote self-deprecatingly to his friend. For those who cared about such things, the reputation of Philadelphia medical schools, like that of its French academies, owed at least as much to the social status of the faculties, students, and trustees as it did to the quality of instruction. The ties between the city's medical establishment and its social elite were well known and frequently noted. The "temper of the trustees at the University of Pennsylvania" was "influenced very much by the power of old families in Philadelphia," a city doctor informed a Virginia friend who was a candidate for a position there. His characterization of the trustees was designed both to encourage his friend to pursue the chair, coming as it did with a certain amount of prestige, and to imply that qualifications alone would not sway the board. Students were certainly sensitive to these concerns. When word reached them in 1819 that the contract of Thomas T. Hewson, an anatomy instructor, would not be renewed, they emphasized his cultivation, not his medical knowledge or teaching abilities, in pleading that he be retained. Their respect for his pedagogical skills, they wrote, was surpassed only by their "unqualified regard of his urbane and gentlemanly manners."[29]

In letters home students stressed the exalted place their schools occupied in the eyes of high-ranking Philadelphians. The implication was that some of this refinement had rubbed off on them. A North Carolina student described the 1825 "grand commencement" at the University of Pennsylvania as featuring "Military officers of rank, . . . respectable citizens & graduates & medical students, & faculty." This "very respectable & imposing procession" marched to a "room filled with ladies & gentlemen," received their degrees, and later attended "an excellent cotillion." Young men did not stress this aspect of their experience to astound provincial kin with their newly won urbanity. This kind of reporting confirmed to anxious parents that their sons were in fact engaged with the refined atmosphere of Philadelphia's medical world. Slave-state critics of northern training did not fully grasp the point when they charged that most of what students learned in Philadelphia would be irrelevant to southern conditions. A Philadelphia education could be social as much as medical, and the standards (and benefits) of gentility were as applicable in Charleston or Natchez as they were in Philadelphia or Saratoga Springs. Young men sought to acquire the qualities that would fit them to participate in a national, not merely regional, upper class.[30]

Learning and credentials helped students qualify for membership in the community of physicians, but young men pursued gentility in other ways also. Illicit activities, including sexual behavior, were hardly respectable pursuits. That was the point, to a degree. Middle-class etiquette advisers from the time of the Revolution excoriated gambling, drinking, and womanizing as the behavior of a corrupt aristocracy. Upright people also stigmatized working-class men as dissolute hooligans, and abolitionists associated slaveholding with these and other vices as well. While "middle-class evangelical and Yankee ideas on this score were different from the mainstream of southern mores," however, middling southerners denounced these vices as vociferously as their northern counterparts. Such behaviors were less southern than aristocratic phenomenon. And while laboring men in northern slums appear to have engaged in whoring, gambling, street fighting, and the like partly in repudiation of middle-class condescension, debauchery among elite medical students should be understood as an important element of upper-class identity, not as evidence of regional distinctiveness.[31]

Sectional ideologues warned that southerners would be corrupted by the allegedly debased moral atmosphere of northern cities, but correspondence between students and their friends reveals that southern men and their northern colleagues shared a common concept of appropriate male behavior. Young southerners anticipated freedom from parental oversight and rural ennui, not from a restrictive regional standard of morality. "When we are at a distance from this place knowing the numerous sources of amusement which it holds out we are left to conceive that we would be as happy as kings," a Virginian confessed to a friend. Southerners faced a period of adjustment when they arrived in Philadelphia. Their unease stemmed from the shift from city to country more than from South to North, however. They quickly learned that Philadelphia, one of the largest, fastest-growing cities in the United States, offered opportunities to engage in youthful vices with neither the trouble nor the guilt of parental oversight. Soon after his arrival, James Brannock of Tennessee confided to his brother George that they had "both, no doubt, caused [their father] to see more trouble than we ought." Medical school provided not the chance to reform his behavior, but to indulge himself without paining his father. "City life is very pleasant to me since I've got used to it, & got the *green* somewhat rubbed off," he explained.[32]

Both northern and southern gentlemen considered themselves entitled to engage in illicit sexual behavior with women whose race and class placed them beyond the pale of respectability. In the South young planter-class men helped themselves to slave women, who were powerless to resist their advances.

Southerners studying in Philadelphia anticipated the opportunities to prey on white working-class women and prostitutes, against whom they displayed similarly misogynistic attitudes. "Since you have been in Philadelphia," one University of Virginia student inquired of a friend studying in Philadelphia, "how do you like [seducing] white gals?" Some students from the slave states experienced difficulty in treating white women with the same cavalier disregard they directed at slaves and free women of color. One North Carolinian looked on jealously at the "presumption, or impudence" of some of his fellow students, the lack of which prevented him from "form[ing] an intimate acquaintance with some of the handsome girls of this city." Most students, it seems, had little trouble viewing poor white women as little more than objects for their pleasure. When a student from Petersburg, Virginia, was accused by a prostitute of being the father of her child, his friend assumed that "being a professed whore would be of sufficient evidence to defeat her villainous intentions towards you in any court of justice." The Philadelphia student suspected that the charges were true, but that was of little matter. That a woman of such degraded status presumed to make such an accusation against a gentleman was itself a shocking breach of social convention. ("I suspect but that she would permit a black man to have intercourse with her," his Virginia friend surmised.)[33]

Young men were far more likely to speak with such frankness to their brothers and friends than to adults. Even though fathers and mentors had perhaps engaged in such behaviors themselves, as responsible adults they could hardly countenance it in their young charges. Nevertheless, some mentors recognized that sexual indulgence was an important part of socialization into the world of male privilege. Josiah Nott, a noted racial theorist who had trained at the University of Pennsylvania, advised James Gage when the young man set out to study in Philadelphia in the 1830s. Nott passed on locations where illicit amusements might be found. The female clerk at a candy shop behind a theater would provide sex "if you manage things properly," Nott advised. Moreover, he intimated that the mention of his name to his former landlady's niece would be "a passport to you, and will admit you to more privileges than you ever dreamed of in your philosophy." Mentors never suggested that students attempt to seduce women with whom they socialized at formal entertainments, and students restricted their advances to women whose race and class marked them as vulnerable and debased.[34]

Student misconduct was not limited to sexuality, however. They also engaged in fighting, drinking, gambling, and other behaviors that reflected a sense that they were exempt from conventional ethical standards. Philadelphia men, like college students everywhere in this period, were renowned for their hard drinking. Robert Nelson of Virginia complained that his fellow students "drink liquor enough to kill a horse much less a human being." Other observers more pointedly stressed the upper-class roots of student dissipation. A North Carolina man who had fallen in with Philadelphia's "*rich* sect" described to a friend the "few days frolick" he had spent with these "wild ones," a bender that left him nearly destitute. Such behavior almost inevitably led to violence, both between students and between students and locals. When a fellow student pulled Thomas Cash's moustache late in 1853, he "promptly hit him one lick," after which "he has been very good to me." Street violence between southerners and city dwellers was a more serious affair. In 1849 one North Carolina student characterized knife fights between students and cabmen as an "almost daily occurrence." Such episodes were so common that in late January 1827 Virginian Robert Peyton saw fit to point out that none of his classmates had "been put in the watch house, nor have they had any combats with the citizens" since the new year. In describing student rowdiness young men struck an ambivalent note. They knew they were betraying parental expectations, but they also reveled in behaviors that violated the canons of middle-class respectability. Common experiences such as these helped mold northern and southern men into a national privileged community.[35]

Dissolute behavior was a rite of passage for young men and a declaration of scorn for middle-class propriety. The ultimate expression of aristocratic aspiration, however, was sociability in select companies of refined women and men. Just as the rounds of balls, teas, parties, and concerts engaged in by young women served as an "informal curriculum" whose lessons reinforced classroom instruction, social affairs complemented the professional education received in medical lecture halls, clinics, and dissection rooms. Attending lectures and surgical demonstrations would not help students acquire the manners and contacts that distinguished them as gentlemen. Well-positioned southern men gained access to the fashionable world via friendships with other students, faculty relationships, parental contacts, and family relations. Gentlemen pursued sociability for its own sake, as a respite from the grueling rigors of student life.

But just as it did for young women, sociability served practical purposes. It cemented relations between privileged families, confirmed one's social rank, facilitated business partnerships, established professional contacts, and might even pave the way for marriages between the most ambitious southern students and Philadelphians of equal social position. These relationships, which spanned generations as well as geography, served as the conduits through which the gentry class passed on its elitist sensibility. The men who in 1815 identified themselves "as Medical Students," "Americans," and "Citizens of the World" to the trustees of the University of Pennsylvania gave voice to this urbane vision. What is more, this ideal transcended mere professional affiliation. It was a fundamental expression of a national, even international, aristocratic identity.[36]

Though families were far more comfortable granting independence to young men than women, the same web of local contacts and relations that made southern parents feel comfortable sending their daughters to Philadelphia also operated with their sons. These contacts of kinship and friendship fulfilled a myriad of important functions both for parents and children. The former could be assured that local acquaintances would monitor boys' academic, moral, and social development. George Hendrickson, an "old friend" of Virginian James Brannock, looked after the Brannock's son during his student years in Philadelphia. "He has treat[ed] me very kindly," the young man reported to his family. "He has invited me around to take tea with his family several times." Mothers and fathers no doubt took comfort in such reports, but not only because they were heartened about their sons' well-being. Chaperoning was a form of hospitality, as young Brannock's invitations to tea suggest. But hospitality was also inseparable from honor; seeking the welcome of others risked rejection of one's claims to social position. It could be a source of pride and affirmation but also of humiliation. Young men's monitors affirmed the rank of the family whose son they looked after. Conversely, the Philadelphians whose aid was entreated had their honor confirmed.[37]

Equally important, parents expected their Philadelphia-area friends to provide access for their sons into polite society. This aspect of their education exposed young men, usually for the first time, to social circles outside their neighborhoods and families. It inaugurated southern students into the world of gentlemen on a far wider scale than they could experience in their local communities. Southern men won access to this society through a variety of channels, the most common being their fellow students who lived in and around the city. Julius Haywood was loath to impose on "a young man by the name of Page" when the Philadelphian invited him on several occasions to his home. "I refused until I was *ashamed* of myself," Haywood confessed, until he "resolv'd at last to brave the battery." He and three other students, all southerners, finally attended a ball at which were invited several Carolinians, a number of locals, and four of Page's sisters. One of these young women left Haywood "thunderstruck," rendering him happy that he had, at last, accepted his friend's offer of hospitality. His father approved for an altogether different reason that speaks to the reasons why some southern parents were so adamant that their sons not limit themselves to academics. A physician's duty, the elder Haywood lectured, "calls him into daily intercourse with all ages, where delicacy & manner are not trivial matters." Such virtues "are only to be well acquired to any practical purpose in the proper cultivation of female society." Local friendships were anything but an incidental benefit of a Philadelphia education. They were a fundamental part of the process by which southern men—and those from rural areas generally—became American gentlemen.[38]

The most urbane students participated in fashionable society not because it gave, in William Haywood's words, "a relish to all the minor enjoyments of our [men's] being," but because they sought initiation into the world of the urban elite. Circulation in polite circles confirmed their high rank and encouraged fellowship with other scientific and medical gentlemen. Socializing inevitably involved fellow physicians—Robert Coleman bragged to his family about his "little circle of professional friends"—but students were most pleased when they were introduced into a wider company. Not everybody could expect to be as fortunate as Daniel Lassiter, who received invitations to formal affairs from Franklin Bache and Phoebe and James Rush (son of Benjamin Rush), but students had other avenues available to them. Abner Grigsby, from an old Virginia family, accepted an invitation from a lady to "take a walk up Chestnut" Street one winter evening, participating in a favorite pastime of the city's establishment. He also attended numerous teas and other functions "by special invitation," as he primly informed his family. Neither young men nor their parents thought of these affairs as diversions from the real business of becoming a doctor. Inculcating the ways of gentlemen was a fundamental aspect of southerners' Philadelphia experiences. John Powell described his life away from class as a process of socialization, through a "quite extensive, but select" acquaintance with his fellow gentlemen, an association that "bears the character of imparting every thing which a young man may desire." Lectures and study provided sufficient training for the mechanics of medical practice but were inadequate for the cultivation of character and contacts.[39]

Professors and wealthy physicians in Philadelphia initiated younger men into their professional fraternity by engaging in sociability with them. The faculties of Philadelphia schools customarily hosted annual parties for their students. Moreover, they held informal affairs throughout the year. In these all-male gatherings, professors and students mingled together casually, eating, conversing, and drinking in a spirit of fellowship. Although many students were intimidated by the older men, the most ambitious among them exploited the opportunity by ingratiating themselves with the professors and other luminaries. In 1850 Clarence Robards attended one such party held at the house of a faculty member. "Seven professors were present, around whom might be seen groups of students," he observed. Abner Grigsby attended a "brilliant party" at which not only "all the profs are present, [but] also some few of the medical brethren." These parties provided access to the community of medical gentlemen for the most striving students. Social affairs communicated lessons about manners, conversation, and connections that would allow young men to maintain and even advance their position within the American establishment.[40]

The atmosphere of these affairs embodied the elitist sensibility of Philadelphia's medical establishment and privileged families. Faculty parties were renowned for their cordiality and extravagance. Informal contacts such as these, in which deference gave way to conviviality, fostered collegial bonds between professors and students. The goal of fellowship transcended professional identity, however. Every glass of madeira, each tray of oysters, demonstrated to students the benefits of membership in this republicanized aristocracy. Dr. George B. Wood dazzled John Owen, a Penn student from Tennessee, with his extravagant hospitality. Owen gushed to his mother that "a servant opened the door and carried us in to a room to take off our cloaks hats & after which another servant came and showed us the way in to the parlor." Wood met his students at the entrance to the parlor, where he "gave [Owen] a hearty shake of the hand and then showed me a seat." At another affair Abner Grigsby observed that the "ice cream vanished far more rapidly than a mist before the rising sun and not one of the huge pyramids of cake were suffered to remain." Liquor flowed freely as well. At Dr. Wood's party servants poured "wine and when [guests] had as much as they wanted of this, the servants brought champaign wine in abundance." Both Owen, who aspired to gentry status, and Grigsby, who was born into it, enjoyed these affairs. Owen sought more than a Philadelphia degree; he craved the urbane manners and national connections that signified gentle rank. And Grigsby found his own position strengthened by association with his fellow gentlemen. Both spurned the pious, moderate, and egalitarian conventions of middle-class respectability in order to embrace the reactionary sensibility of the aristocracy.[41]

Gentlemen craved such connections, both for their professional growth and for the tangible and intangible social perks it bestowed. When David Ramsay, a Philadelphia-trained Charleston doctor, opened up a correspondence between his mentor Benjamin Rush and the Scottish physician Alexander Garden, Rush remarked with pleasure on Ramsay's "progress in reputation &

in the esteem of your fellow citizens" in his adopted city. "I anticipate your rising on some future day to the first honors of your State," Rush rhapsodized. Nearly seventy-five years later, Richard Arnold expressed a nearly identical sentiment when he visited Philadelphia for a conference. While he felt honored to be "amongst so many distinguished medical men," he was also "pleased to find that men who are distinguished in their profession here are not afraid of a little recretation." His companions induced him to leave his beloved opera in the third act to visit a local medical club replete with billiard rooms, fine wines and liquors, and posh accommodations. The evening dispelled his provincial notion that Philadelphia's medical establishment was "all as stiff as if done up in buckram." Though three-quarters of a century separated Ramsay and Arnold, both were men of ambition and cultivation who craved the national recognition for which their professional skills and social position qualified them.[42]

Although a Philadelphia medical education certainly encouraged men to develop a broader, more cosmopolitan sensibility, it was not an obstacle to regional identity. Just as young ladies who had been educated at the city's French schools became Confederate women, Philadelphia-trained doctors served their region proudly during the Civil War. By one count at least five hundred graduates of the University of Pennsylvania served in the Confederate armed forces.[43] Nevertheless, the southern presence in Philadelphia's medical schools built strong and lasting bridges between the sections in the decades before the Civil War, strengthened the city's southern character, and reinforced the cohesion of the American upper class. In 1818 Coleman Rogers, a southwestern doctor, asked that his friend Samuel Brown "remember me to all the professors" at the University of Pennsylvania. Rogers's request was in earnest. His years at Penn had forged a strong sense of personal and professional identity with the city and its medical establishment. "Inform Dr. Chapman that I am in hopes I shall see him again before many years," Rogers asked. "I am much attached to him. He treated me with great hospitality when in Phila." Even after a mighty Civil War, little had changed. On the eve of secession Richard D. Arnold declared "the people of the North are a foreign and hostile people to us and I wish no alliance with them." After four years of war and hardship, however, Arnold reported to a friend that he had recently "stopped four days in Philadelphia & was most kindly & warmly treated by my old friends." Perhaps victory had made the city's medical community feel magnanimous, but perhaps too they offered a heartfelt embrace to an old friend. No longer an enemy, Arnold was one of them once again.[44]

5

Science and Sociability

The American Philosophical Society

THE UNIVERSITY OF PENNSYLVANIA and Jefferson Medical College were not the only Philadelphia institutions that helped nationalize the American gentry in the early years of the nineteenth century. The American Philosophical Society also encouraged conservative men to see Philadelphia as a kind of capital in this period, though its influence waned during the 1840s, the very years when the city's medical schools were strengthening their ties to the slaveholding regions. Although the society's southern ties atrophied for a variety of reasons, for several decades it was an effective proponent of an ideal of genteel scientific endeavor. Men of means, with conservative sensibilities and a strong sense of cosmopolitanism leavened by an intense love of country, discovered in the American Philosophical Society an institution animated by precisely this vision. The society vigorously sought out members who could advance its agenda, and such men sought membership in it. This enthusiasm turned out to be no match for democratic and professional trends that eroded the authority of the amateur man of science in the early nineteenth century. By the 1830s the society's scientific repute was in rapid decline. Yet gentlemen continued to value participation in the society's affairs. Thus, it continued to bind together the nation's elite class well after its scientific authority had atrophied. The power of this sociable ideal proved so powerful that the society's southern ties, though weakened in the years before 1861, did not break until challenged by the Civil War.

The society's standing with southerners and with men of affairs in general was largely the product of a single individual, John Vaughan. As we have seen, Vaughan was a member of the Carolina Row circle, valued as much for his conversational gifts as for his benevolent temper. But Vaughan had an active life outside the company of his Carolina friends. He was a successful wine merchant and business agent. More to the point, he served as the treasurer (1791) and librarian (1803) of the American Philosophical Society, in whose rooms he resided. It is just a slight exaggeration to say that for more than forty years, until his death in 1841, John Vaughan *was* the American Philosophical Society. Certainly this is true for those members who resided outside Philadelphia and communicated with the organization via correspondence. Vaughan dealt with these men personally, sending out subscriptions, soliciting and receiving specimens and articles, and answering inquiries.

When Vaughan died at age eighty-six in late 1841, much of the society's vitality vanished also. He was so closely identified with the fortunes of the American Philosophical Society and the ideals for which it stood that the significance of his death was immediately apparent. William Henry Furness, minister of Philadelphia's First Unitarian Church (which Vaughan attended), observed that the city "is no longer the place it was." Vaughan's death, together with those of several other gentlemen, had altered the city's place in American society beyond recognition. "The very city which they inhabited and honoured had passed away with them," Furness observed. A "new city must be built with new fountains of influence and honour." As the minister suggested by linking Vaughan with such anachronistic values, his passing represented not merely a generational transition, but a cultural shift. Once the provenance of leisured gentlemen meeting in clubs whose purposes were social as much as learned, science was becoming more specialized, academic, and impersonal. Some men socialized in that atmosphere were able to make the transition to the new standards, but most were not. Vaughan's passing was emblematic of another, larger loss, that of a genteel model of scientific endeavor.[1]

Vaughan's death also dealt a blow to Philadelphia's position as the center of gravity for America's privileged families. As treasurer and librarian, Vaughan had succeeded in making the society a center for privileged men from across the young republic. Vaughan used his position not only to connect these men to the society, but to introduce them to each other, forging connections based on similar interests and outlooks. Vaughan also made the society a center of sociability by encouraging men to visit the city, showing them its lions, and introducing them into its fashionable world. "Seldom does a private individual make himself so extensively known," Furness observed. "As we think of the intelligence of his death extending to our cities and villages . . . we can hear every where some one who knew him exclaiming, 'Ah! the kind old man is gone at last. I can never forget him. I rejoice in the remembrance of his friendship.'" Furness was not exaggerating the extent of Vaughan's circle of acquaintance. His influence in the South, in particular, bound that region's learned men to their Philadelphia peers. After his death, links between the society and southern scientists withered.[2]

During Vaughan's long tenure the American Philosophical Society came to embody two fundamental qualities of masculine gentility—learning and sociability. The society emphasized neither quality but combined them into a model of upper-class public engagement. During the early nineteenth century, the society advanced a vision of scientific progress that mirrored conservatives' ideal of social relations. Just as gentlemen believed that ordinary people should defer to their superior abilities in public affairs, they also assumed that scientific endeavors were their natural provenance. Such responsibilities were the price of wealth and privilege, they believed. Also, such men possessed the public spirit to pursue study for the good of society instead of short-term gain. As the society's first *Transactions* explained, "Men of Learning and Enquiry should turn their thoughts to" scientific and technological subjects because ordinary people "seldom turn their thoughts to experiments, and scarcely ever adopt a new measure, until they are well assured of success and advantage from it, or are set upon by those, who have weight and influence with them." The leaders of the society believed that they served the public, but this sense of mission carried with it the assumption that ordinary people should know their place. In fact, talent was often a secondary consideration for membership. The organization "rarely if ever opened its doors to a supplicant who had not already made his mark, if not in science, than in some equally illustrious field such as banking or politics." Because the society's vision of scientific endeavor blended sociability with research, social position weighed as heavily as ability in judging the suitability of candidates.[3]

The American Philosophical Society was certainly not unique among early republican learned societies in restricting its membership to privileged men. Its national reach was distinctive, however, and its attention to establishing bonds to the South was unique. In this respect, the society's efforts complemented those of the University of Pennsylvania and Jefferson Medical College. All three fashioned a nationwide network of learned gentlemen who looked to Philadelphia as their spiritual home. Both the society and the medical schools were animated by a strong sense of nationalism to extend their reach beyond Pennsylvania's borders. Interstate connections would strengthen a sense of American identity, they believed. They would also foster the development of an American science worthy of respect around the world. Thus, the American Philadelphia Society was similar to institutions such as the

Table 5: American Philosophical Society Membership, 1770–1861

Total Members Inducted:	1210
Foreign Members:	393
Pennsylvania Members:	546
Remaining American Members:	271

State	*No. Members*	*Last Year Inducted*	*No. Inducted, 1841–1861*
Alabama	1	1825	
Delaware	9	1844	1
Florida (incl. West Florida)	3	1786	
Georgia	2	1786	
Kentucky	5	1857	2
Louisiana	3	1834	
Maryland	24	1854	5
Mississippi	2	1825	
North Carolina	6	1818	
South Carolina	22	1848	1
Virginia	30	1856	4
Total Southern	107		13
New York	46		15
Massachusetts	43		11

Source: *Proceedings of the American Philosophical Society; Transactions of the American Philosophical Society* (Philadelphia, 1770–1861).

Boston Athenaeum in restricting its membership to wealthy men with similar worldviews. But, unlike that organization, the society's vision was national, not local. It sought out men of means from all areas of the nation (as well as offering honorary memberships to foreigners). And unlike the Boston Athenaeum with its neo-Puritan ethic, the American Philosophical Society followed a more relaxed standard of male sociability.[4]

The society's pre-1841 southern reach can be seen in table 5, which assesses the membership of the institution from its founding until 1861. It is not surprising that Pennsylvanians predominated; from 1770 to 1861, 47 percent of the society's members (67 percent of its American membership) was from its home state. Its cosmopolitan aspirations can be seen in the next largest membership cohort, foreigners. Inductees from outside the United States represented 32 percent of the American Philosophical Society's pre-1861 membership. A substantial minority of the society's domestic membership—33 percent—came from outside Pennsylvania. Forty percent of these men were southerners, with Maryland, Virginia, and South Carolina predominating. Most of the remainder of the American members came from two states: Massachusetts (16 percent of non-Pennsylvanians) and New York (17 percent). The table reveals a clear chronological discrepancy between northern and southern inductions, however. While a third and a quarter, respectively, of New York and Massachusetts members were inducted after John Vaughan's death, only a fraction of southern members were. Only 12 percent of slave-state inductees before the Civil War entered the society between 1841 and 1861. The society's recruitment of southerners was most intense before the era of sectional antagonism, when its nationalizing vision resonated most strongly with urbane southerners. But Vaughan's tenure at the head of the society was also crucial in attracting southern members. Contemporaries recognized that he enjoyed an especially warm relationship with several planting families, particularly South Carolinians. In addition his eighteenth-century sensibility appealed to urbane slave owners who sought friendship with like-minded gentlemen throughout the Atlantic world.

"What [a] great man you are in Charlestown!"

The son of English liberals, John Vaughan was born in 1756. Samuel and Sarah Vaughan were friendly with Benjamin Franklin and other colonial and English republicans. They sympathized with the patriot cause during the American Revolution. John's older brother Benjamin served as private envoy to Sir William Petty, Lord Shelburne, during the peace negotiations. John may have been introduced to the American representatives through his brother; he was already sympathetic to republicanism. He had received some education under the dissenting minister Joseph Priestley. When Priestley immigrated to the United States in 1794, Vaughan was already there. John Jay had introduced him to influential patriots, including the Philadelphia financier Robert Morris, for whom Vaughan came to work in 1782. He went into business for himself after Morris failed in 1784. Morris "was a free liver," Benjamin admonished his younger brother, "not to mention that he was a man of great expense." Whether from Morris's example or not, John Vaughan lived in simple comfort for the remainder of his life in his apartments in Philosophical Hall, adjacent to the State House on Chestnut Street.[5]

By 1787 Vaughan was a member of the society and a successful wine merchant; after moving into Philosophical Hall, he used its basement to store his inventory of wines and liquors. He also served as a business agent for Americans with interests in the Philadelphia port, a group that included a number of prominent southern families. He used his contacts with England and the Continent to supply his clients with books and magazines, news and gossip, and business advice. He supported the ratification of the Constitution and drifted politically toward Federalism, though he took pains to avoid partisan rancor. In her account of her husband, George Logan, a well-known Philadelphia Jeffersonian, Deborah Logan singled out the "benevolent" Vaughan as a "benefit & blessing to Soc[iety]." Vaughan and a few other "federal friends" remained on good terms with the Logans, "kindly visit[ing] & car[ing] for" them in a bitter period of "political excommunication" during which conservative Philadelphians refused all intercourse with the Jeffersonians in their midst.[6]

This universally recognized spirit of benevolence served both the American Philosophical Society and Philadelphia's upper-class community extraordinarily well. Vaughan indefatigably promoted the society and the city. When Jared Sparks, then a tutor at Harvard, visited Philadelphia in 1818, he knew almost no one in the city. But he carried with him a letter of introduction to Vaughan, whom he had never met. The Philadelphian choreographed Sparks's entire visit, introducing him to every mathematician of note in town. After returning to Cambridge, Sparks praised Vaughan to a friend as "the most active member of the Philosophic Society, cicerone and friend to all the strangers who visit the city. . . . recommender-general of . . . every sort of personage, whose characters are good, and who can be benefitted by his aid." His hospitality was responsible for fashioning strong bonds between visiting scientific men and the city. Caleb Forshey wrote to Vaughan from Mississippi in 1841, just before the latter's death, not because he had "anything interesting to communicate, but that I may renew my assurances of gratitude, for kindnesses and attentions I received at your hands when a stranger in Philadelphia" some years before. These good feelings in turn fostered the extension of links between Philadelphia and other regions through the next generation. In his letter Forshey asked Vaughan to keep an eye out for his friend J. W. Monette, a Natchez physician seeking a publisher in the city. Vaughan, Forshey understood from experience, held "the keys to whatever may interest the mind of the inquiring stranger to Philadelphia."[7]

Vaughan's interest in promoting his adopted city was not simply a function of his "benevolence and friendship." Vaughan was also a staunch nationalist.

He cultivated friendships for their own sake, but he also created a web of scientific men to advance the prestige of American science. He was especially interested to bridge the growing differences between the North and South. Joshua Francis Fisher recalled that "in the days . . . when the routes North and South were by the river, he was on the watch for strangers at every arrival, and at once became the welcome for every person of note, or those who came in any way recommended." Vaughan also sought to connect eminent men to the society so that it might advance national interests, as the Royal Society had for Great Britain. Vaughan was "not a man of science or literature," Furness observed, but a promoter and organizer. "He loved to bring men of science and learning acquainted with one another, and to assist all literary investigations, and all sound scientific projects." Vaughan objected to the parsimony of the national government in the early decades of the republic, believing that state support for learned societies was essential for national greatness. In the absence of government largesse, the society depended on the generosity of private gentlemen. As he noted to Jefferson on the Virginian's ascension to the society's presidency, "Pecuniary rewards are not the be had in the present state of Society here, we therefore more strongly require patronage & countenance." Vaughan's determination to make the acquaintance of every man of note passing through Philadelphia stemmed from his desire to promote American science through the support of wealthy, well-placed men.[8]

Vaughan committed himself to promoting the society among learned men in every region of the United States. He enjoyed particularly close relationships with southerners, especially South Carolinians. "What [a] great man you are in Charlestown!" exclaimed José Corrêa da Serra after a fall 1815 visit to the city. "Gentlemen and ladies all asked me about your health, many told me that you ought to pass the winters with them." As the Portuguese priest observed, Vaughan did not limit his acquaintance to scientific men. Rather, he offered a myriad of services to his southern friends that created ties of obligation between the slave states and the city. Vaughan managed Henry Middleton's affairs while he served as minister to Russia during the 1820s. He made purchases for Thomas and Charles Cotesworth Pinckney in the 1790s. When Ralph Izard and his family were away from Philadelphia, either abroad or back in South Carolina, Vaughan looked after their properties. The New Orleans factor of Samuel Brown and Thomas Percy communicated with them through Vaughan when they were in Philadelphia in 1822. Thomas Jefferson used Vaughan as his principal contact with European booksellers when he replenished his library after selling his original collection to the federal government after the War of 1812. Vaughan provided many of these services without charge. When the Mississippi planter William Dunbar looked over his business records after three years of conducting business through Vaughan, he could not "find any charge or commission" exacted by his Philadelphia friend. Dunbar had to plead with Vaughan to accept some remuneration. "I have no right to expect this sacrifice in the line of your regular business," he beseeched. "I must request the favor of you to correct this omission." Joshua Francis Fisher, who knew Vaughan well, remarked that this was his regular practice. To his southern friends, Fisher noted, "he offered more services than it was possible for him to perform."[9]

Whatever his intentions may have been, Vaughan—or more properly the American Philosophical Society—was compensated for his services. In appreciation for his work his southern clients donated materials to the society. As Fisher observed, "if the Society lost in one form (alluding to Vaughan's slipshod accounting practices), it gained more wealth, for he was an indefatigable solicitor for its collection, and gained many presents and bequests in return for his own kind attentions." John Izard Middleton donated a number of rare books, some exotic plants, and "other small matter from S. Carolina" in 1816. His relation George Izard gave the society his copy of a rare botanical work. Henry Middleton sent to Philadelphia a steady stream of Russian curiosities. Joel Poinsett, wishing to "advance by every means in my power the very laudable design" of the society, kept Vaughan busy unpacking plants and minerals from Mexico. Poinsett also directed naturalists from that country to Vaughan's door for hospitality. When James Ferguson sent a collection of poems to the headmistress of a girls' boarding school care of Vaughan, he enclosed an extra copy for his friend, assuming he would prize them since they were "the production of a native of Carolina."[10]

Vaughan possessed a myriad of connections throughout the South, not merely to the leading families of the lowcountry. His friendship with Thomas Cooper, the expatriate English scientist who in 1820 became president of South Carolina College, provided him with connections to the upcountry gentry. Even strangers knew about Vaughan's affection for Carolina, Cooper reported. In 1821 he told his Philadelphia friend that a local gentleman, "whom you do not personally know," heard that Cooper was entertaining Henry Lea, a Philadelphian and a society member. On hearing that Lea was "a friend of yours," the planter offered Lea his hospitality "as *a duty incumbent* on the Gentlemen of S. Carolina toward Mr. Vaughan." Kentuckian Charles Short, the nephew of William Short of the Manigaults' circle, lived in Philadelphia "as a member of [John Vaughan's] family" in his youth. Their relationship created "many obligations" for Short to the society. "Should anything worthy of appearing" in the Society's collections "occur to me or my western friends," he pledged to Vaughan, "I will not be wanting in the will to communicate it." Learned men throughout the slave states felt a similar sense of responsibility toward the society. The Virginia-born gentleman William Preston expressed this sense of regional obligation to Vaughan, showing how individual acts of kindness by the Philadelphia merchant bound the South's learned men to Vaughan and, through him, to the American Philosophical Society. "We of the South are always your debtors," Preston maintained. "I beg of you to give me whenever you can, a chance of [repaying] some of our obligations. When you have a friend going South do sent him to me and if I can [I will] contribute something to your Athenaeum."[11]

". . . drawing the rays of genius to a focus"

Vaughan's personal qualities helped establish a close bond between southern scientists and the American Philosophical Society. His efforts would have amounted to little, however, had planter-scientists not shared his cosmopolitan vision. Learned men throughout the slave states looked to the society as the leading institution of its kind. James Ramsay, son of the Philadelphia-trained Charleston physician David Ramsay, donated a copy of his father's *History of the United States* (1816–17) in 1818 both because it was his father's habit "to present copies of his different works to the Society" and out of "respect for the first Scientific association of our country." But scientific men did not merely see the society as an agent for advancing national prestige. It was their connection to a wider, transnational world of scientific endeavor whose aim was the uplift of humankind generally. National pride, though laudable, should not interfere with the advancement of learning. When Dr. Samuel Brown, then a professor at Transylvania University, procured a skull that he suspected might be akin to the famous mastodon skeleton in Charles Wilson Peale's Philadelphia museum, he sent it to the society so that "some of our learned naturalists will favor the world with a correct description of it." Connection to the society energized southern scientists, many of whom toiled in rural obscurity, with the conviction that their work might contribute to the common endeavor of improving humanity's condition. William Johnson, a Carolina jurist, was so excited on hearing of his election to the society in 1811 that he proposed a new way of voting that would more efficiently integrate the "dispersed" membership. "Much advantage results from drawing the minds of a number of able men," he explained. "It is drawing the rays of genius to a focus."[12]

Because of these connections, Americans in the South and West looked to Philadelphia as a model when they established their own learned societies. Gentleman-scientists such as South Carolina's Stephen Elliot saw in the American Philosophical Society a mirror of their own aspirations. When the Charleston naturalist set out to establish an institution in his city, he appealed to the Philadelphia society to help stock its library and collections. The Philadelphians, Elliot recognized, were well placed "to aid us with our own observations and discoveries." Because of their national and Atlantic connections, the Philadelphia group was also uniquely able to solicit financial support. Elliot asked Vaughan to "forward our purposes" by announcing the founding of the Literary and Philosophical Society of South Carolina "to all within the circle of your acquaintance, whom you may suppose able to render us assistance." Elliott's deference to the older society stemmed not from a sense of regional inferiority, but from a realistic appraisal of relative resources. It also emerged from a commitment to cosmopolitanism that viewed local attachments as parochial. "It is only by the union of many that such institutions can be rendered valuable," Elliott declared. "They are formed to concentrate the diffused and detached fragments of human knowledge, to collect the information which the modesty or indolence of individuals might otherwise be permitted to perish in silence."[13]

The fate of Elliott's society sheds further light on why the American Philosophical Society, rather than regional institutions, remained the lodestar for southern men of science. Charleston's first families proved far more adept at offering encouraging words to Elliott than at contributing funds or attending lectures and demonstrations. "We meet with so little encouragement from our wealthy and fashionable citizens," he complained in 1822, fifteen years after the society's founding, that he doubted it could continue. Six years later, Elliott, desperate to save the society, appealed to Joel Poinsett, the state's most prominent "public man" and one with many ties to Philadelphia, for aid. "I begin to feel anxious as to our prospects," Elliott confided to his friend. Too many members were "irregular in [their] exertions" on the society's behalf. To attract public interest, Elliott sought Poinsett's aid in borrowing a French painting through Joseph Hopkinson, the president of the Pennsylvania Academy of the Fine Arts. One of the academy's paintings had been on display at the Charleston society for more than a year. Its exhibition had helped the society gain "much ground in public opinion." Elliott urged Poinsett to see that when the first painting was returned, "another would be sent in its stead." Southern societies such as Elliott's had difficulty merely surviving. The American Philosophical Society, by contrast, seemed to be thriving with its strong local support among Philadelphia's gentlemen, its international membership, and its national scope. Little wonder that southern men of science, despite their regional pride, identified their interests with the Philadelphia society.[14]

The South's gentleman-scientists remained true to the American Philosophical Society even as it lost influence in the increasingly specialized world of mid-nineteenth-century Euro-American science. The Philadelphia group remained "an eighteenth-century learned society. It had neither the funds, the equipment, nor the organization to contribute much to the developing 'hard' sciences." It also lacked the inclination to do so, which made it doubly attractive to learned southerners. Privileged men had little use for narrow, merit-based, academic science. They sought an institution that confirmed their self-image as cultivated, progressive gentlemen engaged with similar figures around the Atlantic world. Charles Short's ambition to "form a correspondence and exchange of specimens, with almost all of the Botanists of our country, and several of the more eminent in Great Britain and on the Continent," perfectly captured the sociably scientific goals of the society. Vaughan and other officers complied by connecting American scientific men with fellow scientists, particularly traveling researchers. When the French naturalist Louis Gay-Lussac planned a trip to the Southwest in 1838, Vaughan supplied him with letters of introduction to Daniel Drake in Cincinnati, Charles Short in Lexington, Archibald and William Dunbar in Natchez, and William H. Robertson in Mobile. Northern men had stronger institutional networks with which to coordinate their endeavors, but many southerners lacked these urban foundations. Without the American Philosophical Society, they would have been more isolated from the scientific world.[15]

William Dunbar's connection to the society illustrates the benefits of membership for southern scientists. The Scottish-born astronomer immigrated in 1771 to the United States and settled in the wilderness around Natchez, Mississippi, where he carved out a profitable cotton plantation, surveyed the southwestern frontier, and conducted research in various sciences, including botany, zoology, and astronomy. Inducted into the society in 1803, he submitted twelve papers to it between then and his death seven years later. Like other learned men on the western frontier, Dunbar relied on John Vaughan for books, apparatus, journals, and scientific gossip. Dunbar was committed to the society's mission to promote "useful knowledge"; shortly before his death he offered to draft from memory the blueprint of a British spinning machine and have Vaughan contract for its manufacture. Such a machine would be "highly advantageous to us who manufacture all our wool into winter clothing," he explained. In the Old Southwest scientific research could be inextricably bound up with the mundane concerns of day-to-day survival.[16]

The society's facility for supplying Dunbar with scientific apparatus, scholarship, and connections with learned men ranked far higher on the expatriate Scot's scale of priorities than business services. Dunbar relied on Vaughan to put him in contact with Andrew Hamilton, a Philadelphian from whom he hoped to acquire "seeds of plants of the tallow & tea trees" for his botanical garden. After Dunbar acquired a six-foot reflecting telescope "& other valuable instruments," he called on Vaughan to put him in contact with the astronomer Andrew Ellicott, from whom he looked "to profit from the instructions which he is so well qualified to give." Dunbar charged his Philadelphia friend to receive his shipments from London, "instruments books &c," because of the higher shipping and insurance charges on voyages from Britain to New Orleans. In addition to the materials that came to Mississippi through Philadelphia, Dunbar treasured the intellectual community through which he was connected via the American Philosophical Society. Thanking his Philadelphia correspondents for their friendship, he wished that "the most perfect harmony will reign among you, upon which will greatly depend the valuables & discoveries & information expected by the Genl. Govt. & by our Country."[17]

Dunbar's sentiments resonated among men of mind in the rural South, especially in the early decades of the nineteenth century when its frontier regions were still underpopulated and isolated.[18] The avidity with which these men sought out connections with their counterparts across the country testifies to the longing for intellectual community and gentlemanly sociability throughout the South. These men saw the society as a center of intellectual engagement. This often worked better in theory than in practice. The difficulty of communicating between Philadelphia and the rural Southwest strained southerners' connections to the society. These obstacles sometimes led to resignation, but they might also inspire men to redouble their efforts. Caleb Forshey, a Mississippian who had been brought to the society's attention as a man who had "devoted much of his time to scientific pursuits . . . in this remote section of the union," complained in 1841 that he had not "yet received a single number of your *Transactions*" in Natchez, though he had subscribed during a recent trip to Philadelphia. He pleaded with the society's officers to send him the missing *Transactions* and *Proceedings,* pledging that these difficulties had not diminished his commitment to "benevolence and the diffusion of knowledge." Harry Toumlin moved to Mobile, Alabama, in 1822 from the state's upcountry, which he found "absolutely too lonesome," his neighbors consisting of "two or three mulatto families." He appealed to the society to solicit some "intelligent man versed in business" to move to Mobile, improve its social atmosphere, and develop its commercial potential.[19]

Nevertheless, the isolation of the rural Southwest could depress the spirits of men accustomed to engagement with the cultural world. For these men the society might represent less a beacon of refinement than a cruel reminder of their seclusion. Such it was for George Izard, the elder brother of Margaret Manigault. The society welcomed him into its company in 1807, soon after he moved to the Philadelphia area, having "such confidence in your abilities & application as to enjoy the pleasing expectation of receiving great advantage from your zeal & activity." Those memories chafed Izard after he became governor of the Arkansas Territory in 1825. Soon before his death in 1828, he wrote to his old friend and fellow society member Joseph Hopkinson in an effort to recapture a sense of the community to which he had belonged. "I am now living out the remainder of my days among the ignorant, the brutal, the unprincipled," he rued to his old friend. In his last years, he sought a connection to the enlightened world yet again. "I fear to intrude upon your time, which I know to be generally more valuably employed than in perusing epistles from the wilderness," he entreated. "Shall I be disappointed in receiving one from you in return?"[20]

The depth of sophisticated southern men's commitment to the cosmopolitan ideals of the American Philosophical Society is revealed in quite a different way by the example of Thomas Cooper. The expatriate Englishman had been an active participant in the society before he moved to Columbia in 1820. Cooper's "experience & character," wrote Corrêa da Serra to Joel Poinsett, rendered him a "very superior candidate for the position." His supporters were accordingly shocked when Cooper repudiated the society's principles by his violent turn to southern partisanship during the state's nullification crisis. In a much-publicized 1827 speech, Cooper suggested that South Carolina should "calculate the value of [the] union," as if the republic served no greater end than local self-interest. To make matters worse, Cooper's civility seemed to have disappeared along with his cosmopolitanism. Infuriated by a series of antinullification articles in Robert Walsh's *American Quarterly Review,* Cooper sent Vaughan information purporting to show popular support for the movement in the state. "If Mr. Walsh means to deal honestly" with his readers, Cooper demanded, the *Review* would substitute his data for the "misinformation with which his paper has so long lulled them." Cooper's rashness led to his estrangement from the society. Not only were Poinsett,

Vaughan, and Walsh close friends, but all three were staunch nationalists. In 1833 Poinsett had introduced two acquaintances to Vaughan, asking that he present them to Walsh. He expected that their activity on behalf of South Carolina's "Union Party . . . will be a passport to their friendly regard." The society appears to have cut off all personal and professional contact with Cooper because of his activities. "I inquire about you every now and then," Cooper wrote Vaughan in a fence-mending 1836 letter that revealed the learned man's isolation from his old circle of urbane gentlemen. The society had no place in its ranks for an angry sectional firebrand such as Thomas Cooper.[21]

". . . the scientific and literary interests of our dear Country"

The American Philosophical Society was not committed to cosmopolitanism alone; it also embodied nationalism. It sought to advance American science. Its members did not see these goals as contradictory. Humankind would benefit from the promotion of science within the United States. Some members believed that America's republican foundations would make the nation the world's scientific leader. Not only would research be based on merit, not aristocratic connections, but the absence of poverty would provide opportunities for men who in Europe would be consigned to oblivion. The society's men also embraced a nationalist vision that was compatible with cosmopolitanism. Love for one's country (or region) was not inconsistent with love for humankind so long as it did not mutate into zeal, the sin of Thomas Cooper. In seeking members outside Pennsylvania, the American Philosophical Society looked for men animated with a strong desire to promote the interests of their country.

In fact the society sometimes exploited its southern connections in urging men there to join it or to assist its nationalistic endeavors. In 1832 Vaughan reached out to Wade Hampton, a prominent South Carolina planter, to suggest that the slave owner reserve a small portion of his landholdings to experiment in the development of silk culture. "My attachment to Carolina is strong," Vaughan assured Hampton in introducing himself. The recommendation of "some of my Carolina friends" had convinced Vaughan that Hampton was the man to approach. Yet the essence of Vaughan's argument was an appeal to Hampton's patriotism. Peter DuPonceau, the society's leading authority on silk, believed that it had the potential to rival cotton as an export commodity. Vaughan asked Hampton and other "public spirited characters" to use a fraction of their lands as an experiment "for the good of the country." The facilities of the society were insufficient to conduct the experiment, he explained. The advancement of American science required mustering the resources of "*private* enterprise and patriotism," qualities that planters such as Hampton possessed in abundance.[22]

Love of country and the ability to promote its interests was nearly as important a criterion for membership as scientific ability. Farmers might be "intelligent and sensible" and craftsmen "expert and ingenious," but privileged men possessed "fortunes [which] enable them to make experiments, which men of narrow circumstances would not dare to attempt." Thus, the society sought out men whose social position and resources enabled them to promote the nation's scientific development. In selecting the southwestern planter-adventurer Nathaniel Ware for membership in 1823, the society praised him as "an ardent cultivator of the science of Botany, and distinguished by his attainments in various branches of Physics." Actually, the committee overstated Ware's qualifications. But he was wealthy, interested in science, traveled, and well connected within old Federalist circles. Joel R. Poinsett's qualifications were even more impressive than Ware's. The South Carolinian, who was special envoy and then U.S. minister to Mexico in 1822–30, had demonstrated his commitment to the society's ideals by donating Mexican artifacts even before he became a member. Poinsett's generosity demonstrated his "zealous disposition & capacity to promote the scientific & literary interests of our dear Country," DuPonceau claimed. Poinsett was a skilled botanist, but other qualities—his travels throughout Europe and Asia, his public service, Unionism, and his social position—made him an ideal candidate for membership in the society. The emphasis the society placed on social position in considering membership is clear in the cases of southerners Landgon Cheves and William Drayton. The former's reputation was so well known, Stephen DuPonceau stated, that "we deem it superfluous to say anything, by way of recommendation." Similarly, Drayton's patrons forbore giving "any details recommending of him" since the members were "amply acquainted with the merits of this distinguished gentleman."[23]

The Historical and Literary Committee, organized in 1815, was the most ambitious manifestation of the society's nationalizing mission. Its goal was to advance the "moral" sciences "in contradistinction to those which have the material world as their object." It solicited materials on Pennsylvania history in particular, but searched out "as many as possible of the public and private documents scattered in various hands throughout the union" so as to write a "History of America in general." Its chair, the Philadelphia jurist William Tilghman, appealed to the society's members "who reside in different parts of the Union" for materials. However, the committee expressed special interest in the history of the southern parts of the United States. "The Historical Societies of Massachusetts and New York have preserved many important facts and documents," DuPonceau explained to Jefferson. The South's humid climate, rural character, and dearth of libraries and learned societies threatened important documents and artifacts with "oblivion." Virginia lacked "any establishment or institution professing the same objects with those of the committee." To preserve records for the "advantage" of the "future historian of Virginia," the committee solicited Jefferson's aid in directing materials to the society. Not only did the Philadelphia group have many southern members—an advantage neither the New York nor the Massachusetts societies could claim—but as a national organization, adjacent to the upper South, it could be trusted to deal fairly with the slave regions as it fashioned a historical narrative of the nation.[24]

The committee met with some success in seeking out significant documents in southern history. Its first order of business was to collect data on the 1728 expedition to survey the Virginia–North Carolina border, which had been recorded by William Byrd. Eliza Carter Izard, it was reported, thought she had some matter pertaining to the expedition in her private papers. In 1815 "The History of the operation of running the line between Virginia and North Carolina, written by Col. Byrd," was listed among the documents possessed by the committee. DuPonceau, a noted linguist, was especially interested in the cultures of the Indians of the new states and territories of the Southwest. He sought out information from a variety of sources which the American Philosophical Society's long southern arm put within his reach. A southerner passed along the name of Eliza Tunstall of the hamlet of Greenville, Mississippi, who was reputed to be an authority of the region's Indian tongues. Others introduced DuPonceau to military men posted on the Georgia frontier, from whom he learned about the Creeks. He cornered Nathaniel Ware at a party in 1820 and pumped him for several pages worth of information on the Natchez Indians. Other authorities in the slave states eagerly responded to the committee's appeal. In 1817 the committee read letters from "J. Gould being a description of West Florida [in] 1769," and "an account of Tennessee by Mr. Newman [in] 1797."[25]

The committee's work in collecting historically significant materials contributed to the epic effort of nation building in the decades following the War of 1812. This project was inextricably bound up with the development of an upper-class identity among American gentlemen. They believed that only the privileged possessed the leisure, resources, and broadness of mind necessary to carry out such an important endeavor. W. B. Bullock appealed to this sense of mission when asking David Mitchell to answer DuPonceau's inquiries regarding the Creeks. Bullock reminded Mitchell that by assisting the Philadelphian he would be helping a "noble institution" attain one of its "grand objects." The society's commitment to this elitist vision of learned endeavor persisted despite what its officers characterized as the unpromising conditions of American life. European historians benefited from the patronage of aristocratic clients and government funds. By contrast, Tilghman claimed that men "who have sufficient leisure to devote a considerable part of their time to [learning's] acquisition and advancement" were "not very numerous in these states." Not only were men of leisure comparatively rare in the United States, Tilghman complained, but the popular stigmatization of scholarship as aristocratic idleness discouraged gentlemen from engaging in it. The committee declared it to be a small miracle that they had found men "able and willing to aid in the promotion of their objects" from whom they had "derived very important assistance." Men such as Tilghman did not solicit the aid of ordinary or middling Americans only because they considered learned activities outside their ken. Sociability, they believed, was inseparable from scholarship. The circle of learning as defined by the society had to be limited to men of appropriate social rank.[26]

". . . strong aristocratick feeling"

Sociability was central to the mission of the American Philosophical Society. It sought not merely the promotion of science, but of a particular kind of learned inquiry, led and managed by men of wealth and influence. Thus, the society's officers worked assiduously to build relationships between prominent men across the country and throughout the Atlantic world. John Vaughan did his part to fashion a community of learning by holding daily breakfasts at his quarters in the society's building. His closest Philadelphia friends, who enjoyed a standing invitation, could expect to find strangers of note whom Vaughan had welcomed to his table. "I frequently met, at breakfast, distinguished travelers or literary men from other states," J. Francis Fisher recalled. Because his business concerns occupied him for most of the day, Vaughan held his breakfasts at the unheard hour of seven A.M., "two hours before anyone else was up," Jared Sparks exclaimed. Despite the early hour, these meals were always well attended. Vaughan kept the atmosphere cordial and informal, the better to set strangers at ease, promote conversation, and facilitate their entrée into the city's fashionable life. William Dunbar, son of the Mississippi astronomer, knew that these meals were just the place to break in his friend Caleb Forshey to Vaughan's Philadelphia circle. Dunbar asked Vaughan to welcome the West Pointer to his "breakfast table, and thro' you, to the scientific facilities and enjoyments which he so well knows how to appreciate."[27]

To institutionalize the promotion of genteel, masculine sociability, in the early decades of the nineteenth century the American Philosophical Society established a formal mechanism to tie men of affairs to their institution. These affairs, which came to be known as Wistar Parties, integrated far more "strangers" into the city's fashionable world than Vaughan's breakfasts could. They were inspired by Dr. Caspar Wistar, an anatomy professor at the University of Pennsylvania and president of the society from 1815 until his death in 1818. He was well known for combining learning with sociability. "His house was open to men of learning, both citizens and strangers," noted a eulogist. Wistar embodied the gentlemanly virtues of learning, manners, and hospitality. Wistar shared Vaughan's gift for making men with conflicting convictions feel easy in each other's company. "The harmony in which he lived with friends of both parties, and the respect and affection which friends of both parties entertained for him," Tilghman intoned, provided a "memorable example, well worthy of the serious reflection of those who suppose that political intolerance is essential to political integrity."[28]

When Wistar married in 1798, his fellow society members organized an informal club so as not to intrude on his domestic arrangements. Until 1811 they met at his home only on Sunday evenings to drink wine and converse. From then until his death the group imposed more regularity on their affairs, which were becoming overlarge. They limited meetings to Saturday nights during the winter social season, when the first families moved from the country to the city. Wistar served wine, cake, fruits, and ice cream. He sent out about twenty invitations, with guests encouraged to bring prominent strangers who were in town. His most frequent guests were Vaughan, William Short, Joseph Hopkinson, and Nathaniel Chapman. These well-connected men guaranteed that Wistar's parties would feature a "sprightly and gay" company. By offering hospitality to visitors from outside the city, these parties helped establish Philadelphia as a center of upper-class life. The combination of exclusiveness with understated elegance helped establish a counterpoint to nouveau riche vulgarity on the one hand and middle-class simplicity on the other. Wistar's parties "cultivated social feelings, they led to and fostered private friendships, they diffused a spirit of true and elegant, but simple and unambitious hospitality," the association's historian recalled. They acquainted strangers with "the worth, wit, and learning of Philadelphia, and contributed . . . to a just appreciation of the city, its institutions, and its character abroad."[29]

Wistar's friends resolved to organize into an informal association so as to continue the parties after his death. Vaughan, Tilghman, DuPonceau, Robert Walsh, and several others each gave three parties during the social season (reduced to two with the addition of several new members soon after). It was an exclusive group: only society members were eligible to join, and they had to be elected unanimously. To develop the society's influence outside Philadelphia, local membership was limited to twenty, though no limit was placed on the number of "strangers"—men from outside the Philadelphia area—members could bring to parties. It was essential to monitor the qualifications of members closely so that only "eligible" strangers, possessed of features "desireable" to the Wistar Party's aristocratic membership, were invited to its affairs. Conventions dictated that entertainments be kept simple lest "mixed [that is, non-elite] and crowded companies, late and inconvenient hours, [and] sumptuous and extensive banquets" inhibit sociability. Vaughan ran the club until his death, after which the members organized formally, spelling out the by-laws, procedures for the election of officers, and requirements for membership.[30]

Its exclusivity made the Wistar Association an aristocratic anomaly in a republican society. It succeeded in fashioning a sense of community among conservative, learned men not only in the United States, but throughout the Anglo-American world. Henry Singer Keating, a twenty-six-year-old Englishman who spent three weeks of his 1830 visit to America in Philadelphia, was among the "strangers" admitted to a Wistar Party late that fall. He praised their facility for welcoming "all strangers properly introduced to the agreeable society of Philadelphia." The *ton* of these affairs, Keating observed, reflected the "strong aristocratick feeling" that pervaded "the wealthy part of the community." Basil Hall, a Scot whose account bristled with condemnation of America's democratic culture, felt at home at these affairs, where he mixed exclusively with "most of the men of letters, and science, and general information" in the city. Thomas Hamilton, another aristocratic Briton, observed that the parties "bring together men of different classes and pursuits," encouraging the "correction of prejudice." But as Hamilton explained, behind the Wistar Party's hospitality toward middling men lay an elitist purpose. When the privileged gentlemen of the group condescended to invite the occasional "modest and deserving" man to their affairs, it was expected that "his errors are corrected, his ardour is stimulated, his taste improved" by association with the great. Through the Wistar Parties, the American Philosophical Society sought to sustain traditional relations between wealthy patrons and ordinary, though promising, clients, a means of mobility quickly losing relevance in the dynamic social milieu of the mid-nineteenth-century United States. Charles Daubeny, another English guest, captured the aristocratic spirit of these affairs when he characterized the occasional appearance of a middling men as "a wish on the part of the leading citizens to reconcile aristocratic predilections with the exigencies growing out of republican institutions."[31]

The Wistar Association's officers were far more interested in establishing solidarity with aristocratic men from other parts of the Union than they were with middling folk. Hence, they took care to set a civil tone. While conversation was the sole source of entertainment at these affairs, controversial topics were to be avoided. Indeed, many of the society's members held strong opinions on slavery. Several of its Philadelphia members were leading members of the moderate antislavery societies that proliferated in the early republic, though none had any sympathy with the radical movement that emerged after 1831. Civility for gentlemen was more valued than justice for slaves. William Tilghman explained that "considering the position of the southern states, the subject [slavery] is delicate." Slavery's introduction into the Americas was "an event deeply to be lamented," but not acted on: it was enough that "every wise man must wish for its gradual abolition." Because they so treasured civility, and because their censure of slavery explicitly renounced aggressive action to bring it to an end, the society's members were able to accommodate diverse views about the peculiar institution. Charles Caldwell, whose phrenological theories of African inferiority buttressed the proslavery theories of Thomas Cooper, delivered Wistar's eulogy to the Pennsylvania Medical Society even though Wistar had been president of the Pennsylvania Society for Promoting the Abolition of Slavery. John Vaughan, the intimate of numerous slaveholding families, attended the church of the antislavery minister William Furness and assisted Robert Walsh with the distribution of antislavery literature. Defining sociability as a greater good than slavery was an evil, the society maintained the civility of the Wistar Parties until 1860, the very eve of the Civil War.[32]

Just as the hospitable atmosphere of Philadelphia medical schools helps account for the size of the southern-student colony in the antebellum era, the tone established by the Wistar Party made its affairs popular with visiting southerners. Slave owners were frequent guests of members and were even elected to membership. Langdon Cheves, the South Carolina patriarch who served as president of the Second Bank of the United States, was the most conspicuous southern member, having been elected in 1821. Virginians

Robley Dunglison, Nathaniel Chapman, and Moncure Robinson were inducted between 1821 and 1839. In addition, many of the association's Philadelphia members had close family connections with the South. Charles Ingersoll, René LaRoché, and Joshua Francis Fisher all married women from large slave-owning families, through which they maintained contacts to the South. Others had family members or friends in the slave regions, including Vaughan, George Cadwalader, Thomas Wharton, DuPonceau, Robert Hare, Joseph Hopkinson, and Nicholas Biddle.[33]

Planters visiting Philadelphia often found themselves sipping claret and dining on lobster in the parlor of a Wistar Party member on winter evenings. A guest list to Wistar Parties survives from the year 1848, when Nathaniel Beverley Tucker, Joel R. Poinsett, Henry Drayton, and John Bell, among others, enjoyed the association's hospitality. Wistar Parties were part of the itinerary that Vaughan or other local society members plotted for their southern guests. Breakfast at the society's rooms, a meeting of the society, a visit to the Athenaeum, and a Wistar Party—alongside sightseeing or attention to business affairs—might occupy the days a visiting planter spent in Philadelphia. Elizabeth Geffen noted that the visitor's book of the Philadelphia Athenaeum "is filled with the names of southerners whom [Vaughan] took to that organization's reading rooms." When Thomas Percy of Alabama traveled to Philadelphia to place his friend's daughter at a French school, Vaughan invited him to a party, where he met Robert Walsh, "some pleasant men of learning," and a number of foreign diplomats. Both Percy and his northern hosts benefited from this hospitality. Close to Vaughan's friend Samuel Brown, the Alabamian was well read, sociable, and well connected, a pleasant companion for the Philadelphians. Percy, who felt smothered by Alabama's rural ennui, relished cultivated society. The Wistar Party furthered the American Philosophical Society's goal of creating a national network of learned, conservative gentlemen who, like William Gaston, were "habitually joyous and kind in his social intercourse."[34]

Like the society's connections with the South, the Wistar Party atrophied after Vaughan's death. Old rules mandating relatively modest accommodations were ignored; refreshments became extravagant, and the guest lists grew large and unwieldy. In some respects the large companies helped the society maintain civility in an era of mounting sectional tensions. In a diverse group it was possible to avoid those who did not share one's political or social views. Conflict was avoided but at the cost of the Wistar Party's mission to fashion an upper-class community of learned gentlemen from all parts of the nation.

Likewise, as the society inducted fewer members from outside Philadelphia after 1841, the frequency of "strangers" attending these affairs declined.[35]

Yet both the American Philosophical Society and the Wistar Party plodded on after 1841, increasingly irrelevant to upper-class life and American scientific development. Sectional tensions did not actually fracture the social peace in the society until 1861. The trauma caused by Lincoln's election and the South's secession could no longer be ignored for the sake of social peace. As the Wistar Party's historian recalled, the association had forged learned gentlemen into a "community," failing "only when the strains upon that community were too great to be endured." Unionists claimed that the "defiant and outspoken treason" of those who "hobnobbed to the health of President [Jefferson] Davis" was responsible for the Wistar Party's disbanding. Actually, the antagonism cut both ways. In 1861 the group decided to suspend all meetings for the season because of the war. The next year, the same reasons—"only more strongly presented," stated a member—caused the suspension to be extended. A consensus emerged that the club ought not to meet until the conflict was resolved. "The discussion of political questions could not be prevented," George Sharswood observed, "and disagreeable scenes of words if not other kids of collisions might occur." Gentlemen of all political stripes deplored the spreading intolerance that made sociability in a diverse company impossible. Moncure Robinson, a southern sympathizer, resigned from the association in 1861, explaining that "the present unhappy condition of things" rendered him "really unfit for society."[36]

Although the American Philosophical Society's mission to forge American gentlemen into a community of privilege and learning faltered after 1840, its ideal remained attractive to sophisticated planters. In fact, the growing anachronism of the notion of a gentleman as one uniting the virtues of privileged birth, learning, and refined manners may have intensified the society's allure in the eyes of men who still valued that ideal. As we have seen in the examples of sociability as represented by the Fisher family circle and in education and service as represented by Philadelphia's leading medical schools, some men still aspired to it. Hugh Blair Grigsby of Norfolk, Virginia, was such a gentleman. His letter acknowledging his 1857 election to the society shows that this ideal retained its appeal for some planters even on the barren ground of the antebellum period's last years. Grigsby exulted that he would be associated with "the names of those who have earned so enduring a reputation for themselves and for our common country." The Virginian also honored "fellowship with the members of the Society" because of its connections with the South. "The fame of your Society has been ever dear to the people of Virginia," he observed. That the American Philosophical Society no longer represented the vanguard of science mattered little to men such as Grigsby. In fact, its symbol as an anachronistic marriage of science and sociability stood out as a significant merit for gentlemen who were uneasy with American culture's drift toward middle-class respectability.[37]

6

"... ALL THE WORLD IS A CITY"

Travel and Tourism

JOHN IZARD MIDDLETON had nothing on his mind but "pleasure" as he made his way from Charleston to New York early in the summer of 1834. He enjoyed some pleasant diversions in the capital with some gentlemen of leisure, but he spent a considerable amount of time in Philadelphia before he reached his destination. He was held back by a number of family and neighborhood obligations. He dined with Colonel William Drayton and his sons, who had recently relocated from South Carolina to Philadelphia. He called on Harriet Wilcocks, Margaret Manigault's daughter, who had married and settled down in the city. "Liked her and all her family exceedingly," he assured his mother. He spent some time with the Butlers, also close friends of his family, and passed on what gossip he could glean about Pierce's rumored marriage to Frances Anne Kemble. "There is a good deal of mystery about the whole concern," he reported. By mid-June he had put Philadelphia behind him and had arrived in New York. Still, the city had made a strong impression on the young Carolinian. As he told his uncle, Philadelphia was "the only genteel place I have seen since I left Char[leston]."[1]

Middleton was no country bumpkin in awe of the big city. Like many planters, he was a sophisticated gentleman with much experience in cities and towns. Nevertheless, given the well-established links between his family and its circle with Philadelphia, his affinity for the city is hardly surprising. But his response was shared by many other southerners who did not enjoy his local roots. Wealthy Americans had long enjoyed the privilege of leisure travel. But a myriad of transportation improvements facilitating domestic travel after

1815 enabled middling people to engage in these activities on a large scale for the first time. Tourism became an important industry; guidebooks, hotels, stage, canal, and rail routes competed for the resources of casual travelers. Entrepreneurs developed hitherto out-of-the-way attractions such as Niagara Falls and the Virginia Springs to tap into this burgeoning market.[2] These developments benefited Philadelphia and strengthened the city's ties to the slave states. Being adjacent to the South, the city was accessible to southerners in a way that other northern destinations simply were not. Its prosouthern reputation also drew travelers wary of cities such as Boston. However, Philadelphia possessed unique features that made it a popular tourist destination in its own right. Its civic improvements, parks, and historical attractions drew leisure travelers to the city. Well-connected southerners also looked forward to circulating in its polite society. The closeness between Philadelphia and the South, already made strong by ties of commerce, education, kinship, and friendship, was made stronger by travel and tourism.

As with promoters of other sites, Philadelphia's boosters carefully cultivated the city's image as a tourist destination. Travelogues, newspaper and magazine articles, and etiquette books trumpeted the benefits of travel. For individuals travel cultivated the manners and increased one's understanding; its benefits were primarily educational. Eliza Haywood of North Carolina pronounced herself "amply compensated" for her 1824 trip to Philadelphia. Her journey had increased her "knowledge of places and things, which constitutes in my estimation, the substantial advantages of travelling." As a means to culture, travel constituted one part of the national mania for self-improvement that proliferated in the first half of the nineteenth century. Travel, particularly the domestic variety, was also believed to benefit the nation. Exposure to people and places outside one's neighborhood eroded local, parochial attachments and bound Americans more closely to one another. Although it had many critics, this was also a merit of foreign travel. As Virginian Conway Robinson wrote in 1853 after crossing Switzerland, "the tendency of a trip to Europe is to make any reflecting American more and more pleased with the republican institutions of his country, and more and more convinced of the manifold advantages resulting from our glorious Union." Though traveling for pleasure was growing, it was still time consuming and expensive and thus limited to those of substantial means. To encourage leisure travel, local boosters developed the highly constructed practice of tourism, which stressed entertainment as much as education while catering to the time-pressed traveler. One of the many antebellum guidebooks to Philadelphia strove for "concise and amusing rather than learned and minute" descriptions only of the "most remarkable and interesting objects which our city presents to the notice of visitors."[3]

Southerners might establish a sense of identification with Philadelphia by reading about its myriad places of interest. Visiting the city personally endowed those feelings with a special intensity, however. These first-person experiences took on great significance in the late antebellum period, when sectional antagonism caused many southerners to dwell on those features that distinguished them from other Americans rather than those that bound them together. To take one extreme example, the popular southern novelist and editor William Gilmore Simms described himself in 1842 as "an ultra-American" as much as "a born Southron"; ten years later, he had become a rabid sectional partisan. Face-to-face contact between southerners and northerners at resorts such as Saratoga and social capitals such as Philadelphia reinforced a common code of upper-class conduct, maintained and established friendships and family alliances, and displayed the cohesion of the American social elite.[4] The southerners who traveled to Philadelphia were as diverse a cohort as travelers from any other part of the nation—Whig and Democrat, evangelical and agnostic, elite and middle class. But the planters were a distinctive group. Urbane southerners shared with their Philadelphia peers a unity of vision that travel helped reinforce. Privileged Philadelphians and southerners shared far more in common with each other than they did with the ordinary people of their respective sections. The Georgia jurist Richard H. Clark observed that Savannah's first families were "much better acquainted with Boston, New York, and Philadelphia than with our interior towns and counties."[5]

Planter-travelers to Philadelphia tended not only to be urbane, but urban. Like most Americans, southerners viewed cities with some suspicion. They were widely believed to be centers of vice, poverty, and social disorder. Nevertheless, privileged southerners longed for the amenities, cultural contacts, and excitement that only cities could provide. Charleston, Savannah, Natchez, and New Orleans possessed institutions such as libraries, concert halls, and literary and scientific societies that lent the overwhelmingly rural region a significant cultural presence. But planter-class women and men relished the opportunity to visit the great cities of the Northeast, where such resources existed in greater abundance than in the South.[6] Cities penetrated the ennui that, sophisticated people believed, characterized the southern countryside. "There is something that possesses my imagination when I am in Phila. that is a little irregular," Virginian Hugh M. Rose exclaimed in 1825. "Directly that I get in the midst of a city I feel like all the world is a city!" Southern travelers were attracted, not repelled, by the bustling, urban atmosphere of

Philadelphia, just as were medical students, pupils at French schools, and their parents. Northern and southern gentlefolk participated in an "urban culture common to both sections" in the first six decades of the nineteenth century.[7]

Philadelphia was many times larger than most southern cities. By 1860 its population numbered well over half a million souls. This population was also very diverse. Large numbers of German and Irish immigrants called the city home by the 1850s, and a dizzying variety of churches, newspapers, languages, and other cultural markers could be found there.[8] The resulting babel made some planters feel uneasy, but it had the same effect on many Philadelphians. More often, southern travelers felt exhilarated by the city's size and energy. Drury Lacy, a Virginian in town in 1839 to attend a Presbyterian synod, claimed to "see the whole world in miniature by merely looking out" of his fourth-story hotel window. At nearly midnight, he exclaimed, the street was nearly as busy as it had been that afternoon. Elizabeth Ruffin, another Virginian, spent her first day in Philadelphia walking "through many streets viewing their splendor" until her feet were "aching from fatigue." Southerners found the bustle of the streets irresistible. James Henry Hammond came to the city in 1836 to seek medical treatment, yet found himself entranced by the allure of the streets. He had taken several long walks, he reported to his wife, finding that "the streets are so clean, the shops so fine & the houses so beautiful that one never gets tired here." Some visitors found street life to be the most remarkable aspect of their visit, far more memorable than the amusements and attractions that they had come to see. Matilda Hamilton's party was struck "*more,* than any thing else" by the "streets so gay with carriages, omnibuses, & the constant passing of foot passengers." Their engagement with street life left them "so broken down by night, that we all go early to bed."[9]

Three aspects of southerners' experiences in Philadelphia were particularly effective in strengthening their bonds with their northern peers: sightseeing, historical tourism, and sociability. Southerners' presence at attractions such as parks, museums, prisons, and red-light districts illuminates their reactionary social vision via their pursuit of exclusivity and sensual pleasures. They also took in patriotic and historical attractions. They took great pride in their American heritage, identifying simultaneously with their nation and region. Moreover, viewing these sites inspired planters to articulate a peculiar vision of American nationality that exalted privilege, hierarchy, and social harmony as national virtues. Finally, the most eminent planters were able to secure access to Philadelphia's fashionable society. Their hosts welcomed them as peers, not second-class slave owners. Regular travelers keenly felt their exclusion from high society, though both elites and ordinary people testified to the southern atmosphere that made travelers from the slave regions feel at home at the city on the Delaware.

"... the farce of relative republican equality"

Travel guides urged their readers to visit parks, promenades, prisons, workhouses, and other symbols of civic progress. These attractions were, in fact, among the favorite destinations of southern visitors to Philadelphia. Parks and prisons might seem to have little in common, but both affirmed reactionaries' belief in the possibility of limited social progress while, at the same time, promising separation from the urban rabble. Middle-class pedestrians enjoyed the same garden paths and urban retreats but for different reasons. The gentry believed that civic institutions such as the Eastern State Penitentiary and urban idylls such as the Fairmount Water Works confirmed their vision of conservative social evolution. This ideology, which has been called "practical republicanism," combined a belief in regulated economic development with a conviction of privileged peoples' moral and intellectual superiority.[10] They tempered their optimism about social improvement with skepticism about the capacities of ordinary people. Middling people, evangelical Christians in particular, were less prone to impose limits on humankind's potential for improvement. At the same time, conservative folk saw urban attractions in negative terms, as refuges from ordinary people. Both middling and upper-class women and men were plagued by the leers and insults of the working poor. But such people seemed to fling their best insults at genteel folk—a fact that sometimes led Americans in Europe to express admiration for repressive police practices that could not be instituted at home. Middling people enjoyed urban amenities too, but the wellborn patronized them primarily as havens from the company of ordinary folk.[11]

The Eastern State Penitentiary affirmed conservatives' vision of ordered progress and also offered isolation (in extreme form) from working people. The prison was the ultimate expression of upper-class social control, since the dangerous classes were, literally, locked up behind bars. The jail's appeal reflected the gentry's commitment to enlightened advancement with their desire to appreciate sites of aesthetic and civic importance. In describing this and similar places, tour books employed the idiom of "improvement." In the pre–Civil War period this term signified moral and social progress directed by humankind according to the divine plan. This millennial vision was anathema from the world view of the more worldly upper class, who interpreted Philadelphia's public spaces and civic amenities more in terms of social stability and good taste than endless progress. Travel books actually called attention to jails, workhouses, and asylums, but not only because interest in humane reforms was part of the gentry's self-image. Their authors believed that public buildings testified to the urbanity and significance of local elites. Thus, one travel book described the penitentiary as "resembling some baronial castle of the middle ages." The guide paid lip service to the institution's ostensible purposes—incarceration and rehabilitation—but emphasized its function as an object for aesthetic pleasure for privileged women and men. A visit to the prison seldom prompted introspection or guilt in such tourists, many of whom exhibited a myopic complacency about American society. Instead, the Eastern State Penitentiary's combination of architectural style and civic purpose confirmed the gentry's image of a benevolent aristocracy ruling over a quiescent populace. To planter tourists the Eastern State Penitentiary represented the wisdom and virtue of their class.[12]

Completed in 1829 and situated north and west of downtown at Cherry Hill, the jail was a forbidding sight. Its twelve-foot-thick, thirty-foot-high granite walls, imposing battlements, and wretched population were enough to give pause even to the most earnest reformer. Nevertheless, it became a must-see tourist destination. Because it had been inspired by the "panopticon" of Jeremy Bentham, the English utilitarian, the prison epitomized the spirit of enlightened progress to which most cosmopolitan Americans subscribed.[13] The "Pennsylvania System" practiced there combined solitary confinement with industry for the purpose of rehabilitation, not punishment or revenge. Sympathetic observers of the prison could assure themselves that they stood in the vanguard of enlightened reform. Needless to say, in practice the system fell far short of its ideals. Solitary confinement produced a battery of psychological troubles. Moreover, violence, mistreatment, and sexual abuse practiced by guards and officials undercut the system's benevolent designs. Privileged sightseers did not let these gritty details disturb their interpretation of the prison, which confirmed their view of the world and their place within it. J. C. Myers's travel guide improbably portrayed the prison as "situated on one of the most elevated, airy, and delightful sites in the vicinity of Philadelphia." Its aesthetic properties reflected social harmony, not class discontent. Myers's description, with its allusions to "massive square towers . . . embattled parapets . . . [and] pointed arches," evoked a fog-shrouded Byronic ruin situated safely in an imagined medieval past rather than a more disturbing contemporary image, such as the Bastille.[14]

It would be going too far to say that the inmates were irrelevant to tourists, but few were interested in what they might suggest about social unrest or injustice in the early Victorian United States. Visitors could purchase tickets to see the prisoners up close in order to see reformation at work. The inmates, after all, represented an ideal working class—subordinate, harmless, and deferential—qualities that contrasted sharply with those of the actual people genteel folk would soon encounter on the other side of the prison walls. Virginian Matilda Hamilton's visit there in 1857 was little more than another stop on her extended shopping spree. "It is a very nice, orderly looking place," she noted approvingly. "They have solitary confinement, never permitted to speak, or see each other." Before she left, Hamilton displayed her magnanimity by purchasing some of the items the inmates made to sell to visitors. As Hamilton's comments suggest, genteel people were less invested in the transformative potential of incarceration than were middle-class reformers. Rather than viewing convicts as sinners to be welcomed back into the fold, privileged people subscribed to an older reform tradition that sought not to efface poverty, crime, and other manifestations of human depravity, but to ameliorate them. So when visitors ogled inmates like zoo animals and showed more interest in prison architecture than in living conditions, they were certain that they were comporting themselves according to the highest traditions of enlightened humanitarianism. Without a hint of irony, a Virginia traveler characterized an older prison at Broad and Mulberry streets as a "very beautiful and extensive place of confinement."[15]

Ordinary people could not usually be encountered in such restricted circumstances, however. Travelers spent much of their time in public spaces such as streets, shops, and parks, where they were vulnerable to the affronts of working people. The sheer diversity of Philadelphia's population left some travelers feeling uneasy. Henry Massie, a Virginian, took notice of the "motley mixture of sailors, draymen, labourers, & blackguards of every description" he saw on the streets. Gentlepeople observed this population warily. A North Carolinian observed that pedestrians in Philadelphia could not "promenade without the risk of being insulted at every step." To gentlepeople, this represented more than rudeness, which was bad enough. It was a political statement, they believed, one expressing contempt for legitimate social distinctions. Taunts, spitting, splashing mud, and the like were symbolic acts whereby ordinary people expressed their hatred and contempt for their social superiors. Robert Waln, a well-traveled Philadelphia gentleman, believed that the lower classes possessed a "low-bred insolence, and a disposition to insult and abuse those who are their superiors." He yearned for the day when "the aristocracy of fashion and gentility would be more clearly recognized, and the farce of relative republican equality cease to ornament every ragged vagabond with the same attributes as a gentleman." Waln and his peers did not expect their inferiors, as they arrogantly saw them, to be happy about their subordination. They simply wished for public displays of servility. What so upset Waln, Sidney Fisher, and other would-be aristocrats in the first half of the nineteenth century was that few working people even feigned respect for their betters any longer. Insults and physical assaults testified that ordinary folk no longer feared the gentry's power, yet resented their pretensions.[16]

Upper-class women believed themselves to be favorite targets of predatory working men. "I well know it requires great exertions to Deal with the common class," a Philadelphia woman empathized to her Virginia cousin. "They are disposed to Cavel and give trouble to our sex when in many instances they would not have courage to contend with their own sex." Likewise, when some male friends teased Harriet Manigault for becoming distressed at hearing their voices while on a walk, she explained that working men who encountered women "without a gentleman in the country, are very apt to be rude." Perhaps because women's fashions were conspicuous signs of class distinction, ordinary folk really did single them out for abuse. Whether their vulnerability was real or imagined, however, visiting gentlewomen displayed a special affection for urban refuges such as parks, walks, and promenades where encounters with the lower orders could be regulated, if they could seldom be avoided altogether. Jane Caroline North, a Charleston belle, felt safe on Chestnut Street, the city's most fashionable thoroughfare, when she visited in the spring of 1850. For all its élan, however, it was a public street, not the exclusive domain of the gentry. In the midst of a pleasant stroll a "little miserable of a boy" barreled past her, and before she "recovered the shock, he had torn the lace of my defenceless mantilla."[17]

In the early nineteenth century municipal authorities built parks and other urban retreats that brought the country to the city. These improvements were often thinly disguised attempts to segregate genteel people from their social inferiors.[18] Travel guides took pains to point out the quiet squares and high-class residential districts where run-ins with undesirable elements might be minimized. Philadelphia had a well-developed park system, owing partly to William Penn's original design for the city. It envisioned wide, perpendicular streets encompassing four squares set aside as wooded parks, with a central square at the intersection of Broad and Market streets. The middle class enjoyed these havens from the underclass as much as the gentry, but—as the favored recipients of insults and other signs of disrespect—the elite tended to flock to them as much for their isolation as for their promise of aesthetic enjoyment. No other site in Philadelphia combined the virtues of the refuge with the social ideal of ordered progress better than the Fairmount Water

Works. Designed by Benjamin Henry Latrobe in the 1790s and reconstructed by Frederick Graff between 1811 and 1822, the works supplied the city with water from the Schuylkill River. Graff's hydraulic system became the model for water-supply methods in nearly forty American cities. His mechanical marvel attracted many visitors interested in the work's scientific apparatus. Even before the nineteenth-century improvements, a visiting southern physician declared the water works to be "a Grand display of human ingenuity." The works were surrounded by manicured grounds to attract fashionable pedestrians, creating an oasis of refinement in the midst of the bustling antebellum city. As one local guidebook noted, the works "have extensive grounds attached to them which afford a delightful promenade and place of resort." The Fairmount Water Works seemed to affirm the vision of moral improvement that animated conservative women and men in the antebellum years.[19]

Travel books described the works as the perfect marriage between science and aesthetics. J. C. Myers's travelogue, a book popular with southerners, suggested that the works "present an eminent combination of elegance and utility. The grounds are adorned with beautiful walks . . . beautifully ornamented with shade trees of the choicest species." Such descriptions paralleled the cautiously progressive world view of sophisticated southerners. The comments of these travelers reveal that they interpreted the works much as the guidebooks suggested they ought. Jane Caroline North confided to her diary that the "Fairmount works repaid us amply for the fatigue & trouble of going through the dreadful dust" of the dry August streets. Rich and well-educated, the young Carolinian instinctively translated sights and sounds into the language of refinement. The water works, she wrote, were situated on "a rugged steep rock covered in many places with luxuriant vines . . . the green of the long drooping branches mingling with & seen thro' the spray of the fountain is charming to the eye." North made the usual observations about the "immense" machinery, but the "beautiful union of nature and art" really captured her fancy. This fusion was visible via the "small temple[s]" shielding from cultivated eyes the pumps and pipes, the fountains of nymphs and river gods, the whole scene evoking a sense of rustic serenity far removed from the grime of city life. As Albert Jefress, a Virginia Methodist teaching Sunday school in Philadelphia in 1838 observed, the works afforded "shade & relaxation to those who go hither from the noise & distraction of the city."[20]

Both middle- and upper-class Americans enjoyed the serenity of public spaces such as the works. Yet, privileged people desired not only to segregate themselves from the working poor, but from the respectable middling ranks as well. The social critic Fanny Fern praised the Fairmount Water Works for their egalitarian qualities. The estates visible from the river, she wrote, were less enjoyed by their proprietors "than by the industrious artisan, who, reprieved from his day's toil, stands gazing at them with his wife and children, and inhaling the breeze, of which, God be thanked, the rich man has no monopoly." The rich may not have wished to corner the market on breezes, but they certainly did seek to separate themselves as much as possible from the less fortunate. As Fern recognized, they could do so only imperfectly at public sites like the Water Works.[21]

Happily, Philadelphia provided leisure-class southerners a variety of places where exclusivity was assured. The Continental Hotel told prospective guests that "a Vertical Railway . . . extending from the ground floor to the top of the house" provided direct access to the "upper rooms—which have always been regarded as most desirable." The escalator was just one of many features designed to ensure "entire exclusiveness" from the hoi polloi occupying the lower floors. Likewise, certain tourist sites were reserved to the privileged by virtue of their wealth, connections, and distinguished airs. Arriving at Pratt's Gardens without a ticket, which had to be purchased at a downtown office, John Strobia and his party introduced themselves as "strangers" and gentlemen. After paying a "trifling fee" to the gardener, the party secured a private tour. "Every information [was] given us that we required," he wrote smugly. Not only did the gardener personally direct their expedition, but he apologized for having asked this refined company to follow the rules established for ordinary people. The gardener explained that, "having received much damage from the depredations of boys and others . . . the proprietor determined, at last, that no person should enter it without tickets of admission." One the party identified themselves as Virginia gentlemen, they received the special treatment to which, they assumed, they were entitled.[22]

Other places followed a similar policy, appealing to cultivated people and discouraging the patronage of regular citizens. Charles Wilson Peale exhibited his mastery of the common touch in the layout of his museum of natural history specimens and historical artifacts. For comic effect he placed the skeleton of a mouse below the feet of his celebrated mastodon skeleton. Yet his collection remained a favorite attraction among cultivated visitors for decades, partly because the Peales extended special hospitality to them. When Dr. Adam Alexander visited the museum in 1801, he received "friendly attention" from none other than Rembrandt Peale himself. The Virginia doctor could easily afford the cost of admission, but the young Peale "would take nothing from me or my friends." Alexander exhibited the combination of learning, urbanity, and manners that marked a man of stature. His museum tour, during which

Peale "shew[ed] us everything," was an act of courtesy to a fellow gentleman. A commercial transaction would only taint an expression of sociability between peers.[23]

Privileged travelers took in other types of amusement that, while less innocent and by no means exclusive, were no less indicative of the class-embedded nature of leisure travel. Drinking, whoring, and gambling were popular diversions among upper-class visitors, though their illicit nature has left them underreported. To be sure aristocratic practices such as "the eighteenth-century custom of wealthy men drinking themselves under the table" were in sharp decline in the early nineteenth century, the advent of the Victorian era. Nevertheless, even in moderated form, the gentry's diversions looked downright hedonistic next to the stiff propriety of the middle-class ethic. Opportunities for illicit activity Philadelphia and other cities were strong inducements for southern men, especially young ones free of adult supervision.[24]

A pamphlet characterizing "the city of brotherly love" as also being "the city of sisterly affection" was aimed directly at those travelers who had more than museums and tree-lined walks on their agendas. The author sought to protect "stranger[s] . . . against the possibility of being involuntarily seduced to visit a low pest house" during their stay in Philadelphia. As the pamphlet makes clear, even the pursuit of illicit sex was bound up in Americans' obsession with class and social station. The tract ranked houses of assignation according to the status of their clientele. It also paid close attention to the refinement of the prostitutes themselves. Hence, "Miss Sarah Turner" won praise for being "a perfect Queen." Not only was she cultivated and discreet, but her "young ladies are beautiful and accomplished; they will at any time amuse you with a tune on the piano, or use their melodious voices to drive dull care away." It assured visitors, "none but gentlemen visit this Paradise of Love." Yet, as with other aspects of social rank in this fluid society, appearances could be misleading. Mary Spicer's house was "well furnished" and her prostitutes "dress[ed] well." Yet, the tract warned prospective clients away with a bold "X" by her address. Her house's refinement was all "appearances," it warned; its gentility was akin to that staple of antebellum fiction, the confidence man.[25] Another proprietor, "Sal Boyer, alias Dutch Sal," put on no such appearances. "This is the lowest house in the city," strangers were warned. "No gentleman ever visits this Sodom."[26]

Direct evidence for travelers' patronage is scanty—they did not describe these encounters in letters to mothers and wives back home, and diaries were rarely private and confidential—but it is likely that solicitation of prostitutes, gambling, and other forms of dissipation were widely engaged in by southern men traveling in Philadelphia. Court records from New York City examined by Christine Stansell suggest not only that genteel men felt entitled to use working-class women sexually, but that the women's efforts to resist such objectification was seen by authorities as a breach of propriety. Occasionally, however, isolated rays of evidence penetrate the gloom. Some medical students confided to their friends and brothers about their escapades, as we have seen. In the journal he kept of his 1849 visit to Philadelphia, Harvey Washington Walter of Holly Springs, Mississippi, wrote of how he "renewed many of my old acquaintances," with whom he "engaged in some wildness." In a guarded letter to his mother Andrew Polk related his Christmas Eve 1842 escapades on Chestnut Street. From the Delaware to the Schuylkill rivers, the Princeton student explained, the street was "one condensed mass of human beings of all ages, sexes, sizes, and conditions." The public houses offered "'egg nog' parties gratis. We had a superabundance at the Jones Hotel." Everyone on the street competed "with his neighbor in point of happiness and hilarity," he observed innocently. Polk's description is inoffensive enough, but the young man probably omitted the whole truth. Antebellum college students were renowned for their lack of restraint. Far from home, an anonymous man in a gay crowd, it is unlikely that Polk checked the impulses in which so many others of his station felt entitled to indulge.[27]

" . . . what reverence we should feel for those great men!"

Innocent and not-so-innocent amenities drew their share of southerners to Philadelphia. The city had other attractions, however, first among them its places of historical significance. It is easy to emphasize those features of the antebellum South that made it distinctive from other regions of the United States, particularly the Northeast. In fact the South possessed a number of qualities that set it apart, not only in national context, but in the western world generally. In the antebellum period the South was the largest slave society in the world; Britain's 1833 abolition of slavery intensified the South's feelings of marginality. Thoughtful observers—northerners, southerners, and Europeans alike—frequently remarked on the economic, social, and cultural gulf that seemed to distinguish the South from the North and West. Our knowledge that the era in which these differences grew ended in a great civil war seems to justify an emphasis on the distinctiveness of the slave states. To many contemporaries, however—particularly gentlepeople—regional differences seemed less significant compared to the bonds of history, kinship, and culture that bound the North and South together. One reason why southern sectionalism appears so striking is because it developed during a period of intense nationalism. Both northern and southern social elites were committed to the survival of the Union their ancestors had made together. The gentry's national vision helped it avoid the "narrow provincialism" that, so many visitors to American shores believed, marred the young nation's popular culture.[28]

Visiting southerners and their Philadelphia peers developed a peculiar brand of American nationalism that—as in the case of the Charleston physician David Ramsay—sought to "replace an imperial with an American cosmopolitanism without wholly rejecting British culture or withdrawing into a national shell." In addition, planters managed to develop this vision without rejecting or compromising their regional identity. Only sectional zealots had difficulty being simultaneously southerners and Americans. In visiting Philadelphia's historical attractions, privileged southerners did not subordinate regional identity in their national pride. But if their nationalism embraced northerners and southerners equally, in other ways it was highly discriminatory. It slighted, if it did not ignore completely, the contributions of ordinary people. Moreover, this nationalism allowed no room for class-based conflict or division. In paying homage to the historical sites around Philadelphia, planters constructed a relentlessly political conception of American national character that excluded groups they deemed of no account.[29]

Leisure travel did not itself signify a commitment to nationalism, but it usually implied at least benign curiosity about the world beyond one's neighborhood. Travel guides not only reinforced and encouraged nationalistic sentiments, but did so in a manner designed to appeal to their urbane, largely Whiggish readership. *A Traveller's Tour through the United States,* a game designed to relieve the monotony of long-distance travel, was rife with patriotic themes. Players moved the game pieces around a board consisting of a map of the United States by describing the characteristics or history of the place on which their die roll landed them. The game's manual described the United States as "by far the finest portion of the western continent . . . with respect to wealth, fertility, civilization, and refinement." Its portrayal of local characteristics reflected the conservative Whig idealization of the Union as well as their ideal of gradual, regulated improvement. A *Traveler's Tour* depicted Philadelphia as a "noble city. . . . the centre of a great trade," with "the most extensive manufactures of any city in the Union." It also warmly complimented the South while situating it firmly within the Union. The game described Virginians as distinguished for "valour and patriotism." More than just pleasant diversions, games such as *A Traveler's Tour* promoted a conservative, nationalist sensibility in their players.[30]

Like games, travel books fused information with patriotic commentary. They were hardly uncritical, however. Their accounts of urban life fed conservatives' fears about lower-class disorder and the perils of democracy. J. C. Myers's *Fashionable Tour* explained that Pennsylvanians were "distinguished for their habits of order, industry, and frugality." The state's vistas offered the pleasing contrast of "noble roads and public works, with the well cultivated fields." But Myers's travelogue warned readers frankly about the hostility they would confront from ordinary people. Much as they did with prisons and other social blemishes, however, tour books urged their readers not to interpret these signs as indications of fundamental American flaws. "It would be impossible," Myers's guide assured travelers, "to find a like number of cities . . . whose average moral, social, and intellectual condition stand so high." Tour books demanded that readers take pride in their nation, but they defined that community so narrowly that all but the privileged were excluded. Roads, canals, and other features of "progress" were signs of the wisdom of the nation's refined citizens. Seen through the lens of the would-be aristocracy, these developments testified to the wisdom of the gentry's concept of gradual progress rather than pell-mell individualism, rural malaise, or sectionalism.[31]

Visiting the historical sites around Philadelphia intensified and refined planters' national attachments. Personal contact with historical treasures, particularly those central to the nation's founding, lent a palpable sense of connection to American history far stronger than one acquired by study or recitation. Common people from all regions imbued feelings of nationalism from sources such as elections and their concomitant rituals, Fourth of July orations, popular literature, and other sources. By contrast, leisure travel fostered personal contacts with distant people, allowing tourists to engage with history firsthand. Such contact rendered the gentry's sense of national identity personal and concrete. Travel guides reinforced these sentiments by emphasizing the national significance of Philadelphia's local history. One pointed out that "the city is noted for several events in our history," of which it singled out William Penn's negotiations with various Indian nations, the Continental Congresses, and the British occupation during the American Revolution.[32]

These were events to swell the pride of all Americans, not just Philadelphians or those from the mid-Atlantic states. Guidebooks also infused seemingly local attractions with national importance. J. C. Myers's book reminded its readers that the American Philosophical Society was much more than a local institution, pointing out that several of its members were heroes of the revolutionary struggle, including "Benjamin Franklin, David Rittenhouse, [and] Thomas Jefferson." The resonance of travel-based national associations was largely reserved to women and men of means. Few ordinary people could afford the expense of leisure travel. A sense of national pride was hardly limited to the wealthy, of course, but their identification with the nation was more visceral than that of less fortunate people, who were connected to others mainly through imaginative bonds fostered by newspapers, histories, and orations. Mary Telfair, a Georgia gentlewoman, complained that the plain folk of Savannah had "such strong prejudices and [were] so local in their feelings." The well-traveled Telfair noted that her "habits, views, tastes, feelings have all be changed by Northern association." For most southerners, local affairs—inevitably more pressing and relevant—matured into a strong regional identification in the late years of the antebellum period. It was precisely lower-status whites who seemed most "southern" to observers before the Civil War. Leisure travel both reflected and reinforced the southern gentry's nationalist sensibility.[33]

The appeal of Charles Wilson Peale's museum illustrates planters' peculiar nationalist sentiments. Besides his well-known assortment of natural-history materials, Peale possessed a collection of portraits of revolutionary leaders. This feature impressed Virginian John Strobia more forcefully than any other collection in the museum. Peale had accumulated likenesses of "all the leading men concerned in the American Revolution," marveled Strobia, including "Washington, Fayette, Baron Steuben, Greene, Montgomery, Jay; and many other distinguished characters." The portraits' historic significance impressed the Virginia diarist far more than Peale's artistic mastery. Properly interpreted, the Revolutionary era would serve as the foundation for a growing, thriving, and stable American state. "This group, a century hence, will be a valuable collection in the eyes of posterity," he enthused. In creating and exalting such artifacts, privileged people such as Strobia and Peale sought to invent a tradition for the American Revolution. They sought to wrest control of the past before its interpretation could be shaped by those—Thomas Paine, for example—who might take it in more radical direction. It was a conservative, consensus-driven version of events that ignored or, at best, downplayed the disorder and contentiousness of the revolutionary struggle. Reactionaries fashioned a highly selective narrative of the founding era that insisted on the wisdom of the Federalist vision to which, as conservative Whigs, they stood as inheritors.[34]

Strobia wrote early in the century, when sectional tension was at a low level. But planters' intense feelings did not diminish at midcentury, when sectional antagonism was well-developed and on the increase. Even at this point, sophisticated southerners discerned no conflict between regional pride and nationalism. Emma and Annie Shannon were both proud southerners and Unionists (their father was the editor of the *Vicksburg Whig*) when they attended St.

Mary's Hall in Burlington, New Jersey, in 1858. While touring Philadelphia with their schoolmates in the spring, she and her sister were introduced to a "privileged person" who offered to "take us around and show us the places of interest." Carpenters' Hall, whose "ancient . . . red and black bricks" they associated with the epic age of the struggle for independence, rekindled their patriotism. Entering, they "stood in the hall where the first congress was held, the spot where Patrick Henry poured forth his spirit-stirring elegance, &c." The Shannons clearly felt no contradiction between their southern heritage and their national loyalty. In fact, southerners' veneration of the symbols of American nationhood often went well beyond that of their Philadelphia hosts. To visitors, locals seemed indifferent to their stewardship of national treasures such as Independence Hall, which they were bound to maintain for their fellow Americans. Southerners interpreted this apathy as a moral failing on the part of their northern counterparts. A North Carolina visitor in the 1830s complained that the State House "stands unnoticed and unhonored" when it "should be the boast of every Philadelphian." Upper-class southerners saw Revolutionary relics as the "dearest proofs of our freedoms," the most potent symbols of the American identity they prized.[35]

At the same time, southerners were ambivalent about the meaning of sites such as the State House. Independence Hall evoked feelings of both unease and decline, reflecting the ways in which class and nationalism intersected in conservatives' consciousness in the antebellum period. The Revolution was not, after all, a distant memory in the first half of the nineteenth century. Moreover, the French Revolution—as well as uprisings in Europe throughout the first half of the century, particularly in 1848—made it seem more relevant than ever. Its legacy inspired considerable contention. Political parties, ethnic groups, and even private societies and interest groups claimed to embody the true spirit of the American Revolution against their opponents. After 1800, for example, Federalists refused to participate in Fourth of July celebrations administered by the Democratic-Republican opposition lest their presence lend legitimacy to their foes. Both Federalists and planter-travelers of the nineteenth century believed that the individualistic, democratic society of the antebellum period was the product of misinterpreted revolutionary republicanism. Defining themselves as the trustees of the nation, conservatives appropriated symbols such as the State House in order to fashion their own interpretation of the Revolution and its meaning. The real American Revolution, they argued, had been fought in the pursuit of conservative republicanism; its meaning had been perverted to legitimize the democratic dissipation of contemporary times, they contended.[36]

Guidebooks reinforced this selective memory by describing the State House as the embodiment of a deferential past. Patriotic rhetoric contrasted the crass, fragmented, middle-class democracy of the Jacksonian period with the harmony, unity, and clarity of Federalist America. One tour book suggested that visitors interpret the "bell used on the memorable occasion" of independence as a knell "calling the people together." The Liberty Bell was "a relic of the heroic age of American history" that united all Americans. Tourists' ruminations adhered closely to the guidebooks' recommended interpretations. Such feelings possessed Virginian Matilda Hamilton, who visited Independence Hall in 1857. She assumed the State House would be "the most interesting place in Phil[adelphia] to all true Americans." It represented to her the "heroic age" of revolution, independence, and nation building. Visiting the site inspired Hugh Rose to imagine "our father patriots s[itting] to deliberate on the fate of the nation . . . what reverence we should feel for those great men!" From wherever they hailed, privileged Americans believed that the relics of Revolutionary history enshrined an integrated, cultivated, and aristocratic republic.[37]

Planters' reverence for the patriotic sites of the founding era did not imply indifference or hostility toward southern identity—to the contrary. Sectionalism and nationalism were parallel developments that emerged out of the impulse to define the identity of the republic as well as a larger, transatlantic movement to celebrate local peculiarities rather than universal qualities. Staunch southern partisans were not awed or repelled by Independence Hall. The meanings they took from it were not markedly different from those of more conventional southern travelers. The State House hearkened back to a time when sectional differences were submerged beneath life-and-death struggles to win national independence. The edifice tapped into a repressed core of American nationalism even among those who publicly reviled the North. It inspired both regional extremists and nationalist planters to appreciate their place in the Union. Clement Clay, a member of the fiercely sectional Knights of the Golden Circle, was profoundly moved by viewing Independence Hall. When he "struck the old cracked bell [and] sat in the chair occupied by John Hancock," Clay "felt [his] patriotism grow warmer and pulse beat quicker." Visiting the State House was a pilgrimage, a ritual through which visitors transcended differences of creed and section and confirmed their place in a national community.[38]

"The Philadelphians are very distant with strangers"

Unlike sightseeing and historical tourism, sociability was not a common pastime among southern travelers to Philadelphia. Some travelers, aware of the city's southern orientation, were surprised and put off by the aloofness of locals. Philadelphians were known for their exclusivity, but their standards were based on class, not section. Those visitors with the requisite local contacts, letters of introduction, or celebrity penetrated the guardians of high society, but their numbers were few. William Chambers, an English traveler, ascribed the cosmopolitan tone of the city's beau monde to its "happy blending of the industrial habits of the North with the social habits of the South." Chambers missed the point. Sophisticated Philadelphians prized their close relations with their counterparts from the slave regions, but because they were fellow gentlepeople, not southerners. When William Brisland visited Philadelphia on business in 1839, he registered at his hotel as a resident of Natchez because the city stood "very fair" in the eyes of Philadelphians. But he "took tea with Mrs. Ralston" not because he was a Mississippian, but because he was an old friend of her family. Similarly, southerners appreciated the congenial atmosphere of Philadelphia, where they were unlikely to be subject to insult because of their slave owning. But they also enjoyed the city because its dynamism was such a counterpoint to the torpidity of the rural South. As Mary Telfair noted, she had "two characters—a northern and a southern one," the former cultivated and sociable, the latter domestic and reserved. Northern society, she was happy to claim, had turned her into a "dissipated creature, luxuriating on fine scenery and talk." Socializing between Philadelphians and the elite class of visiting southerners further bound them into an upper-class community in the decades before the Civil War.[39]

Travelers were warned against expecting to be welcomed into Philadelphians' parlors and ballrooms merely because they were southerners. Private societies such as the Athenaeum and the American Philosophical Society opened their doors to strangers only under very restricted circumstances. Usually the sponsorship of a local notable or a reliable letter of reference was required for admission, unless the aspirant was so renowned that no introduction was necessary. Whatever the case, strangers who desired access to these and similar institutions needed to demonstrate their gentry bona fides. Travel books described exclusive societies flatteringly but cautioned their readers against expecting to be admitted unless they had a local champion. The Athenaeum admitted outsiders, G. M. Davison's guide explained, only on their "being introduced by a subscriber." Another book praised this society for "furnishing a place of resort for persons of leisure who may wish to read the newspapers, reviews, and scientific journals." Visitors were warned, however, that "strangers" could only be "introduced by subscribers or stockholders." Exclusivity such as this involved more than mere snobbishness, though that was important. Limiting access to private clubs allowed gentlemen to believe that they still exercised considerable power in American society. Exclusivity also helped the gentry isolate their way of life from the corrosive influence of the egalitarian popular culture of pre–Civil War America.[40]

Some travelers, of course, did enjoy the stature to win entrée into Philadelphia's fashionable society. Eliza Haywood of Raleigh was the beneficiary of her father's friendship with the South Carolina grandee Langdon Cheves, president of the Second Bank of the United States. When she visited the city in 1824, the planter himself "conducted us to his new and magnificent mansion, the splendor & comforts of which I could not have conceived without having seen them." Privileged Philadelphians had to monitor access to their society closely, lest they lose their reputation for exclusivity by being too open. Aspirants either had to possess proof of their status in the form of letters or personal contacts or, less likely, demonstrate their cultivation in person. Recognizing that Philadelphians were known for being "cold and reserved in their intercourse with strangers," one southerner noted that those "who bring letters of introduction, or persons whose family, education, and manners are such as to entitle them to move in their circles will, when acquainted, have the most marked attentions paid to them." In fact, Philadelphians were remarkably hospitable—to the right people. "There is no city in the Union in which a gentleman is better received," this writer insisted. Yet only after "pass[ing] the ordeal" of offering proofs of high station would a gentleman be "safe and happy in their society."[41]

The benefits of passing such an ordeal were commensurate with its difficulty. At the most basic level sociability alleviated the loneliness and boredom of travel to a strange city. John Houston Bills, a Tennessee planter, anticipated just such relief while visiting the city in 1845. But finding "many of my friends absent," Bills could only note tersely, "Phila. more dull than I had expected." More fortunate travelers could enjoy a good meal and friendly company while catching up on news and gossip. During his 1859 visit to Philadelphia to enroll his son in medical school, the Mississippi naturalist Benjamin L. C. Wailes took in the usual tourist sights—the Laurel Hill Cemetery, the Fairmount Water Works, and the Academy of Natural Sciences among them. But he found the greatest pleasure in renewing his friendship with the Pease family, with whom he spent an afternoon that "sped away most pleasantly; questions were asked and answered after Mississippi friends, and old occurrences, which a mutual interest hallowed, were discussed." Cultivated southerners also craved the extravagance of fashionable society and the honor that acceptance into it conferred. Middle-class critics may have derided sumptuous entertainments as a source of the gentry's "indolence and perpetually increasing incapacity," but gentlepeople saw dissipated behavior as a class entitlement. Anticipating a trip to Philadelphia in 1823, William Gaston told his friend Joseph Hopkinson that "to mingle familiarly with the delightful society of your city, with the learned and the gay and the polite, is among the highest gratifications which my fancy can conceive."[42]

Most southerners, however, lacked the contacts to enter into such refined circles. Those who were unsure of their credentials were urged not to try, as uncertainly probably meant unsuitability. These people were advised "not to attempt the purchase," since they "will almost certainly fail" to secure entry, one writer warned. This isolation proved to be source of acute disappointment to some visitors. Preachy Grattan, a Virginia divine visiting Philadelphia in 1837 to participate in the Presbyterian synod, became keenly aware that his middling status limited his access to desirable sites in the city. As he complained to his wife, "the fact that I was a total stranger in Philadelphia & had no person to go with me to see any thing or even direct me how to set about attaining admission to the various objects of curiosity which abound in Philadelphia, rendered my visit much less interesting than it might otherwise have been." Philadelphians' vigilance in monitoring access to their society testifies to the importance of class and status in this period of American history. On a more personal level failure to gain entrée constituted a painful reminder of just where one stood on the American social scale. Henry Massie, a Virginia traveler, discovered this unpleasant truth during his 1808 visit. "The Philadelphians are very distant with strangers," he noted bitterly, "but much the reverse, I'm told, with those they know."[43]

Travel to Philadelphia strengthened the bonds between the city and the South, but Philadelphia was also significant for its national associations, particularly its places of historical significance. In that sense, planter travel to the city, perhaps more than the educational or purely social links discussed in earlier chapters, encouraged the development of a national identity among southerners. These bonds were remarkably strong. They were not powerful enough, needless to say, to resist the tide of sectional strife that erupted into the Civil War. They were sufficiently strong, however, to provoke a profound conflict within Philadelphia as that city's social elite contemplated engaging in hostilities with their planter friends and relations. Some gentlepeople openly supported the South, even going so far as to advocate Philadelphia's secession. Most, however, were repelled by the course taken by the now-Confederate states and repudiated their former associations. As sectional tension gave way to sectional warfare, Philadelphians confronted not only dividing their beloved Union, but sundering their own social world.

Epilogue

THE WARM RELATIONS that had prevailed between planters and their Philadelphia friends during the first five decades of the antebellum period persisted during the strained decade of the 1850s and even into the Civil War. The series of slavery-related crises that occurred in the 1850s introduced considerable tension in the relationship, however. Friendships were tested; civility threatened to break down; and minor disagreements flared into serious breaches. Yet even secession itself did not break the ties that bound privileged southerners, Philadelphians, and their peers elsewhere in the United States into an American leisure class. Instead, it was the outbreak of armed hostilities in April 1861 that produced the fatal breach. The Civil War not only severed the relations that bound Philadelphia to the slave South, however. It also had far-reaching effects within the city's social circles. A significant fraction of the city's establishment openly supported the Confederacy during the war. Their behavior discredited them in the eyes of their fellow gentlepeople. Unionists responded by redefining the criteria for membership in the elite class. Because extravagance and prosouthern attitudes were linked, they spurned both (in the case of the former, temporarily). The city's postbellum elite was proudly Republican in character. Antisouthernerism, forged in the heat of local battles for control of high society, was among its defining elements. Those who remained loyal to the old ways retreated into memory, biding their time until that day when, they hoped, elitism would rise again.[1]

". . . the South, whose sympathies are ours"

For a number of reasons Philadelphia's upper crust resisted the antisouthern drift of northern opinion during the 1850s. Racism was ubiquitous throughout the North in the first half of the nineteenth century, but it was particularly virulent in Philadelphia. Frederick Douglass, who knew of what he spoke, averred that "there is not perhaps anywhere to be found a city in which prejudice against color is more rampant than in Philadelphia." Antiblack prejudice was not only to be found in predictable places such as Irish neighborhoods or among the community of southern medical students. The city's upper ranks also expressed racist sentiments with a virulence that was anything but refined. Joseph Willson, a Georgia-born African American Philadelphian, attempted to counter this prejudice in his *Sketches of the Higher Classes of Colored Society in Philadelphia* (1841) by showing that the community of free blacks had "men of fortune and gentlemen of leisure." It was to no avail. The gentry's racism was too ingrained to be susceptible to argument. Charles Brewster, a prominent Democrat, articulated the sentiments of many in fashionable circles when he denied that African Americans were capable of attaining civilization. According to Brewster, Africans were little "above the condition . . . of the baboon that chatters in the tree above him," and blacks in America were little better. Such attitudes created a presumption of virtue toward slave owners among the upper ranks of Philadelphia society.[2]

Such widespread, open hatred went well beyond the insidious but more subtle brand of racism more common elsewhere in the North, particularly New England.[3] Philadelphia's opinion of abolitionism remained particularly low during the decade. In late 1859 the antislavery activist Susan Lesley had to ask a Boston friend for a newspaper so she could read a sympathetic account of John Brown's execution because she could not find one in a local paper. "It cannot be that Boston is so cold as Philadelphia," she wrote hopefully. "Here we are accustomed to people being still in the dark ages." Abolition was still regarded throughout the North as a fringe reform, as dangerous for its endorsement of racial equality as for the threat it posed to the stability of the Union. Its image softened among many northerners during the 1850s, however. White southerners' response to a number of events, particularly the caning of Charles Sumner by Preston Brooks, impelled northerners to take a fresh look at what abolitionists had long been saying about the moral influence of slavery. Philadelphians were no exception. As we have seen, Sidney George Fisher grew increasingly critical toward the South in the 1850s. Nevertheless, this sentiment had made less headway in Philadelphia than in other parts of the North. At a dinner in his honor in 1860, Alexander McClure, a Republican Party leader, admonished speakers to "avoid offensive expressions" against the South because he knew that "Philadelphia was so conservative on the sectional issue."[4]

As secessionist sentiment grew in the South during the last half of the 1850s, conservative Philadelphians reassured southerners of their friendly intentions. Philadelphians strained to point out that few northerners sympathized with abolitionist radicalism. Arthur Ritchie sought to disabuse his Georgia cousins of their conviction that all northerners were antislavery activists. "We have a few fanatics here, *as you have also in the South,*" he maintained. "The majority of the northern people have just as good feelings for the South as they ever had." Elizabeth Grimball also struggled to correct her brothers' warped view of northern opinion when she spent the fall of 1860 at the home of her aunt, Gabriella Butler. Her fire-eating brothers were having none of it. Philadelphians were "cool enough" not to disclose their antislavery views, her brother Lewis wrote condescendingly. To reveal themselves as "the enemies of slavery" would provoke the "dissolution of what they are pleased to call this 'glorious union.'" Other Philadelphians stoked southern hysteria while assuring planters of the loyalty of the city's upper classes. "There seems to be a studied design to put *you* and *us* in the wrong," William Bradford Reed wrote to a South Carolina correspondent in early 1861. Yet he assured his slave-owning friends that his circle of Philadelphia gentlemen was working tirelessly for sectional peace.[5]

Reed accurately represented the behavior of Philadelphia's politically active upper class. During the 1850s they enthusiastically endorsed the South's demands for the preservation and extension of slavery. Adopting the labels "Union" and "conservative" to tar their opponents, prosouthern Philadelphians sought to align their city ever closer to the slave states. During a "Great Union Meeting" (a Democratic Party rally) in the midst of the furor over the Compromise of 1850, George M. Dallas admonished northerners to "rekindle the almost extinguished confidence and friendship of our Southern brethren" by demonstrating that they "sincerely sympathize in the sufferings and wrongs to which they have been subjected." As Dallas's words indicate, Philadelphia conservatives subscribed to an interpretation of sectional relations that saw the South as the long-suffering victim of northern fanaticism. Seeing the Democratic Party as the surest guarantor of southern rights, many gentlemen—William B. Reed among them—abandoned the faltering Whigs for the Democrats in the mid-1850s. These attachments only grew stronger as the decade progressed. At another Union meeting, called after John Brown's raid into Virginia, the directors assured "their Southern brethren" that "patriotic and conservative" Philadelphians were even more upset at Brown's "display of fanaticism" than they were. "Do not mistake the sentiments of a few with the sentiments of the masses," Henry M. Fuller, a speaker at the meeting, pleaded.[6]

As events in 1860 and 1861 were to show, Fuller's accounting of the "masses" and the "few" suffered from serious flaws. Philadelphia's conservative elite may have grown even more prosouthern, but the city and the state were drifting away from the South and its political champion, the Democratic Party. That party, closely identified with the Pennsylvania-born president, James Buchanan, lost control of the state in the 1858 elections. Owing mainly to the large population of Irish immigrants in Philadelphia, the state's nativist Know-Nothing Party was among the strongest in the country. Between 1856 and 1860, the nascent Republican Party successfully co-opted nativist positions, blurring their antislavery credentials. The party further obscured its slavery plank by renaming itself the People's Party in 1860, confounding the opposition's effort to brand Lincoln supporters as "Black" Republicans. Moreover, local Democrats, like the national party, divided. Some favored John C. Breckenridge, while others backed Stephen Douglas. Breckenridge supporters inflamed sectional tensions in the state, arguing that Pennsylvania's "true interests" lay with "the South and Southwest from which . . . Abolitionism, if triumphant, will forever divide us." Such rhetoric backfired; Breckenridge defeated Douglas in the city, but Lincoln carried not only Pennsylvania, but Philadelphia as well, earning 52 percent of the vote.[7]

In the three months following Lincoln's election in November, the seven states of the Deep South seceded from United States and established the Confederate States of America. During that period dozens of meetings were held in Philadelphia to express sympathy with the South and to demand that the government make concessions to entice the states to return. Others urged northerners to accept secession, and some prodded Pennsylvania to follow the Deep South's example. Conservatives pointed out the cultural affinities between the middle states and the South, as when a January 1861 gathering expressed solidarity with "the South, whose sympathies are ours." Others argued that separation from the slave states would result in economic disaster for Pennsylvania. One secessionist urged Philadelphians to imagine "an endless chain, freighted with comforts and conveniences," stretching from Long Island Sound to Cape Fear in a prospective confederacy that excluded New England. The *Palmetto Flag,* a secessionist newspaper, predicted that when the city's industrialists realized that they might "become the manufacturing hand of the Southern confederacy" they would repudiate "the abolitionists and their allies the Republicans" to join the new Confederacy. It was also argued that Pennsylvania's secession would foster peace, as when a writer suggested that it would tip "the scale against [Lincoln], and *therefore against Civil War.*" However, conservatives undermined their cause. Misreading the popular mood, they employed vulgar and unpatriotic language wholly inappropriate for the serious times. At a rally early in 1861, Charles Brewster asked why Philadelphians "should butcher our southern brethren because the Yankee is in love with a he-nigger?" George Wharton's speech came close to treason. "Our interests are with the South," he averred. "Mine are at any rate. The south and the west are our best friends."[8]

These conservatives had badly misapprehended the drift of public opinion. Such attitudes were tolerable during the tense but peaceful winter of 1860–61. Even the city's establishment was sharply divided about the correct policy to take toward the seceding states. If some advocated passivity, others endorsed a hard line. Joseph Ingersoll did so sadly. He reminisced about the times when southerners made "visits of health and recreation" to the North, when "hospitable doors were everywhere thrown open to them." But Ingersoll recognized that those times were gone. The "Union must coerce the South if everything else failed," he concluded. Secession was "too venomous to be pitied, and too violent and mischievous to be despised." Ingersoll's sentiments became conventional wisdom among his set only after the attack on Fort Sumter. The assault transformed the alignment of Philadelphia's fashionable community. On hearing the news of the fort's surrender, mobs surrounded the houses of known southern sympathizes and forced the *Palmetto Flag* to cease publication. Conservatives made patriotic speeches to save their homes from the torch, but others seem to have had a genuine change of heart. Sarah Butler Wister, who was Pierce Butler's daughter, was shocked to find Hartman Kuhn, "despite his Southern sympathies and Loco Foco tendencies," a born-again Unionist. "Patriotism, & feeling for the Union supersede everything else now."[9]

A southern orientation, hitherto a badge of respectability, suddenly became a mark of disgrace. Recrimination and suspicion became routine, making routine civility between Unionist and southern sympathizers impossible. When William B. Reed claimed to renounce secession, his peers tarred him with the alleged disloyalty of his grandfather, who had, it was rumored, betrayed George Washington in private letters while serving as his aide during the Revolution. Wister suspected that Reed's newly discovered Unionism was owing to his "hereditary halter, that noose which has been hanging over his treacherous race for three generations." Even the Philadelphia Club, whose members eschewed serious subjects such as politics in favor of sociability, found it impossible to maintain civility. By 1861 relations between Unionists and southern sympathizers had grown so strained that they occupied two separate rooms. Sarah's son, novelist Owen Wister, recalled that when in 1861, a Union man remarked that the "place reeks of" Copperheads (as pro-Confederate northerners were called), one of those gentlemen promptly "knocked him down and was therefore expelled." The prosouthern consensus that had maintained social peace in Philadelphia since the early years of the republic had broken down. "We live in an intolerant and prescriptive community," observed Peter McCall. "It is no longer the city of Brotherly Love."[10]

"We have the example of the French Revolution"

The dilemma of the Fisher family epitomized that faced by other southern-leaning gentry families during the years of the Civil War and Reconstruction. During the 1850s, it will be recalled, Joshua Francis Fisher struggled to mediate between his increasingly antisouthern cousin Sidney and his fire-eating brother-in-law, Williams Middleton. Williams mocked Fisher for failing to realize that the logical outcome of his antidemocratic convictions was secession—opposition to slavery being, in his mind, a product of democratic excess. Because they valued each other's friendship (and family peace) they simply avoided discussing political issues. Fisher even tried to maintain a middle position after South Carolina's secession. He could not countenance the dissolution of the Union, but neither could he support what he believed to be an abolitionist-inspired administration. When he told Williams that he felt compelled to remain loyal to the government despite his personal feelings, his brother-in-law mocked him mercilessly. "Such resolves on the part of the educated & respectable & of such men as yourself & friends will only the more surely result in disaster to you all," he charged. To try to shape public opinion, Fisher published a pamphlet lamely recommending the repeal of state laws interfering with the fugitive-slave law, a maximum tariff rate of 20 percent, and other measures that would have placated no one, North or South.[11]

Hostilities put an end to Joshua Fisher's ambivalence. The war forced him and Eliza to make a choice, and they chose the South. By late 1861, when it was clear that hostilities would not end quickly, they had become ardent supporters of the Confederacy. Three influences seem to have settled their minds. First, Eliza was in Charleston with her children at the time of Fort Sumter's shelling. What she saw inspired her with the righteousness of the Confederate cause. "I cannot but admire the noble spirit of self-sacrifice which animates" South Carolinians, she wrote her husband. "It seems to be felt universally." Second, Fisher became convinced that the war was not merely the product of, but a catalyst for, democratization. He took special umbrage at Lincoln's speech of July 4, 1861, which characterized the war as "a people's contest." Fisher believed the president sought to expand what was fundamentally a political crisis into a pitiless ideological struggle that could be won only by destroying the South. "How can one expect the South to yield unless they are thoroughly *crushed?*" he wondered. Third, rumors of the Middleton family's suffering, confirmed by letters received from Williams in 1865, impelled them to see the Union cause as a vindictive struggle to break the planter class. Williams told of how his wife and children had barely escaped Columbia before it was razed by General William Tecumseh Sherman's troops. Williams's slaves forsook his "protection" for freedom in the Yankee columns. Worse, the former master found himself a fugitive behind enemy lines. "I, with three or four friends, was hunted like a wild beast," he told his sister. Newspapers and personal accounts fed the Fishers' rage. "My heart is bleeding for our friends at the South," he told his cousin, "reduced from comfort to *starvation.*" What made their situation intolerable was the abandonment of the South by its former friends in Philadelphia.[12]

Fisher and other Copperheads marginalized themselves by their open support for their country's defeat. As the war dragged on and family after family became touched by tragedy, such attitudes could no longer be tolerated. George Fanhestock, who had been friendly with the prosouthern gentry before the war, nearly exploded with indignation during the nervous days after Lee's invasion of Pennsylvania in the summer of 1863, when he saw groups of Copperheads, including Reed, Charles Ingersoll, and George Wharton "radiant with joy." In the context of universal sacrifice, such behavior was not merely tactless—it was counterproductive. Joshua Fisher's conduct also seemed designed to alienate his neighbors. He "absolutely raves incoherently," Sidney observed after a run-in with his cousin in 1863. As Confederate defeat seemed inevitable, however, the Fishers mellowed. Noting his cousin's temperament after a May 1865 social call, Sidney surmised that "the logic of recent events has no doubt had its effect on his mind."[13]

But the Fishers had not been jolted out of their consternation by an acceptance of Union victory. Rather, the Middletons' distress in the helter-skelter days surrounding the demise of the Confederacy rekindled their sense of duty. They shook themselves out of their self-pity on hearing stories of the burning and looting of Middleton Place and of the deaths of friends and family in the Confederate ranks. Realizing that Eliza's family depended on them, the Fishers set out to restore the confidence and economic foundation of the Middleton clan. In 1864 Eliza tried to comfort her nephew Bentivoglio Middleton, serving in the Marion Artillery and Signal Corps, in language that anticipated the sentimental rhetoric of the "Lost Cause" later in the century. "Their noble deeds sanctify the soil upon which they rest," she assured him after noting several deaths in their circle. The couple made every effort to aid their southern relations during the war, including intervening to secure the release of Nathaniel R. Middleton's son from a Union prison. The Fishers, Joshua in particular, were also inspired to act by the degradation of the Middleton family. Williams's loss of mastery, that sense of authority at the core of the planter-class's identity, deeply angered his brother-in-law. No one in the North, he complained to the Fishers in 1865, could appreciate how the war had overturned "our institutions, rules, regulations, habits & opinions & indeed everything." He was humiliated by his inability to protect his family and by the betrayal of his slaves. Williams was also humbled by the responsibility he felt for leading the South into a war whose "result . . . was to bring ruin to four of five millions of whites, misery to almost an equal number of blacks, & loss of liberty to a whole continent full of human beings." The image of the noble Williams brought low filled them with purpose in the last years of the war and the early years of Reconstruction.[14]

But the Fishers were not undamaged by the war. The changes it wrought in their family arrangements, community, and in their sense of place in history inflicted deep psychological shocks on both husband and wife. Feeling betrayed by and alienated from their social circle and nation, Eliza and Joshua responded quite differently. As if to defy their new circumstances, Eliza hurled herself into Philadelphia society. Over several weeks in 1868, Fisher told Williams, she had given several parties, attended some operas and concerts, and spent much time shopping. But he could not reenter polite society, he told Williams. "If she has enjoyed all this, I certainly have not." Fisher was feeling so alienated from his neighbors that he feigned illness to avoid attending a party given by Eliza at their house. Some planter-class women responded to defeat by indulging in excesses—lavish parties, spending sprees, and other conduct entirely inappropriate to the circumstances. It was an act of defiance against failure and humiliation, "an assertion of class privilege in the face—and in defiance—of its rapid erosion."[15]

Fisher responded to personal disappointment and the decline of his class quite differently. He accepted his diminished social position. Withdrawing from society was his expression of resignation. But he also did something more constructive: he dedicated himself to preserving the memory of privileged society in antebellum times. Family history, the practice of memory, was a political act. As he explained to Williams, it was their responsibility to leave to their children an account "of the times before the Deluge, and teach them the principles they ought to hold & the noble ends they ought to aim at." He understood—as, for quite different reasons, Frederick Douglass did also—that memory is not a private act or a mere invention of the imagination. Public memory is a struggle between rival interpretations of history with relevance for contemporary contests. Fisher's recollections represented his effort to forge a weapon to be used in a future struggle he anticipated being waged between competing versions of American history.[16]

Memory was all Fisher had left. The Confederacy had lost the war, and Philadelphia society had been reorganized on new foundation: bourgeois values, Unionist sentiment, and Republican Party membership. But even if Fisher believed that memory was all he had, he thought his children might have more. Privilege had reigned before; perhaps it would rise again. By the time his children reached their majority, "the political tornado we are passing may have spent its force," he hoped. The democratic revolution of the Civil War might have its Thermidor. "We have the example of the French Revolution," he explained to his brother-in-law, "and I am not without hope that I may live to know the names of Sumner & Stanton & Wade as much excoriated as Danton's [and] Robespierre's." For Fisher the decline of the planter class and the American gentry were inseparable. Reminiscing about a gentlewoman from times past who led "a cortege of beaux through Watering places . . . with a body of gay fellows behind her, and keeping the whole party alive by wit," Fisher observed that such conduct was unthinkable under the postbellum era's Victorian moral code. "I only mean to cite this lady as proof of what women in those days dared do—while the great families of this country still retained their prestige." He wished not merely to remind his children of their southern roots, but to provide them with role models for what he hoped would be a future resurgence of aristocratic culture.[17]

The Civil War left Philadelphia's prosouthern gentry with little except the memories and vague hopes for the future that sustained Joshua Francis Fisher in the last years of his life. (He died in 1873.) During the war, Unionist gentlemen established the foundation for a new social order that repudiated the southern orientation and aristocratic aspirations of the antebellum establishment. The institutional embodiment for this new order was the Union League, a private patriotic society founded in 1862 to support the policies of the Lincoln administration. Because "the disloyal talkers lorded it in society," Unionists were forced to establish their own club. To isolate the old guard, it was decided that "disloyal men should be positively excluded from the meetings of the" new organization. As Union victory and their own behavior eroded the legitimacy of the pro-Confederate element, the Union League became the arbiter of social respectability in Philadelphia. The rise of the league "had a powerful influence on the social position of disloyal men," the league's history observed. By drawing clear lines between the disloyal and loyal, and by staking the latter closely to support for the war aims of the

Republican administration—including the abolition of slavery—the Union League helped break the southern orientation of the highest echelon of Philadelphia's fashionable society.[18]

Privileged Unionists also understood the aristocratic aspirations of the prosouthern element of society. Planters and their Philadelphia friends "acquired a notion of their innate superiority to the middle-class rank and file of the Eastern trading folk," Republicans believed. The Union League therefore set the tone for postbellum society by repudiating extravagance and other signs of aristocratic aspiration. The rules established at the first meeting of the Union Club in November 1862 banned all but "moderate entertainment," imposing a three-dish limit and limiting wines to sherry and madeira. These prescriptions, unlike the ban on prosouthern attitudes, had a limited effect—Gilded Age society was not known for its parsimony. The larger point, however, was the repudiation of elitism that the Civil War effected in Philadelphia society. As George Lathrop, an early member of the league, noted, the influence of southern-leaning gentlefolk "from that hour began to wane and never recovered." With the defeat of the South and its local allies, "the old standards passed away, and society in Philadelphia was no doubt materially changed and liberalized." Lathrop observed that the city's antebellum "society had been ruled by rigorous distinctions, often arbitrary, but entirely irreversible; and those who had made the distinctions were in general Southern in their leanings." Such was not the case after 1865. The Civil War gave conservative Philadelphians a reason to fashion a Lost Cause of their own to match that of the planters they so deeply admired.[19]

Notes

Abbreviations

ADAH	Alabama Department of Archives and History, Montgomery
APS	American Philosophical Society
DCL	Department of Special Collections, Dickinson College Library, Carlisle, Pennsylvania
Duke	Rare Book, Manuscript, and Special Collections Library, Duke University, Durham, North Carolina
FHL	Special Collections, Filson Historical Library, Louisville, Kentucky
GDHA	Georgia Department of History and Archives, Atlanta
HSP	Historical Society of Pennsylvania, Philadelphia
LC	Library of Congress, Washington, D.C.
LCP	Library Company of Philadelphia
MCV	Special Collections and Archives, Tompkins-McCaw Library, Medical College of Virginia, Richmond
MDAH	Mississippi Department of Archives and History, Jackson
Penn	University of Pennsylvania Archives and Records Center, Philadelphia
SCHS	South Carolina Historical Society, Charleston
SCL	South Caroliniana Library, University of South Carolina, Columbia
SHC	Southern Historical Collection, Wilson Library, University of North Carolina, Chapel Hill
VHS	Virginia Historical Society

Introduction

1. Baptist, *Creating an Old South,* 5.

2. Morrison, *Slavery and the American West,* 1–13, presents a brief summary of the main interpretations. For a strong statement of this argument, see McPherson, "Antebellum Southern Exceptionalism."

3. Taylor, *Cavalier and Yankee;* Grant, *North Over South;* Horsman, *Race and Manifest Destiny,* 164–68.

4. Waln, *The Hermit in Philadelphia,* 80.

5. An argument made most forcefully by Wood, *The Radicalism of the American Revolution.*

6. Bushman, *The Refinement of America;* Hemphill, *Bowing to Necessities.*

7. Beckert, *The Monied Metropolis,* 10.

8. Pessen, *Riches, Class, and Power,* stresses wealth at the expense of culture and conduct; more sophisticated analyses include Blumin, *Emergence of the Middle Class;* Beckert, *The Monied Metropolis;* Baltzell, *Philadelphia Gentlemen;* Johnston, "The Caste and Class of the Urban Form of Historic Philadelphia"; Burt, *The Perennial Philadelphians;* and Jaher, *The Urban Establishment.*

9. Hamilton, *Men and Manners in America,* 388–89; Leland, *Memoirs,* 136.

10. "Extracts from the Letters of Correspondents," *Southern Literary Messenger* 1 (Feb. 1835), 322–23.

Chapter 1: The Carolina Row

1. King, *Lily,* 74.

2. On the family's background, see Crouse, "The Manigault Family of South Carolina, 1685–1783."

3. Alice Izard to Margaret Manigault, Mar. 31, 1811, Manigault Family Papers (SCL).

4. Zuckerman, "Tocqueville, Turner, and Turds."

5. Wood, *The Radicalism of the American Revolution.*

6. Allgor, *Parlor Politics;* Branson, *These Fiery Frenchified Dames;* Kierner, *Beyond the Household;* Wood, "'One Woman So Dangerous to Public Morals.'"

7. Waldstreicher, *In the Midst of Perpetual Fetes;* Fischer, *The Revolution in American Conservatism.*

8. Alice Izard to Margaret Manigault, Dec. 7, 1808, Manigault Family Papers (SCL).

9. Webber, ed., "Josiah Smith's Diary," 146, 209. Wallace, *History of South Carolina,* II:227; Bell, *Major Butler's Legacy,* 37–38. For a contrary account, see the comments of George Grieve quoted in Chastellux, *Travels in North America,* ed. Rice, II:593.

10. Rasmussen, "Democratic Environment–Aristocratic Aspiration"; Griswold, *The Republican Court.*

11. Bingham quoted in Earl E. Lewis, "Anne Willing Bingham," in James et al., eds., *Notable American Women, 1607–1950,* I:146–47; on Graeme, see Shields, *Civil Tongues and Polite Letters,* 124.

12. George Izard Autobiography (SCHS); Ralph Izard to Peter Manigault, July 1, 1764, Ralph Izard Papers (SCL); Butterfield, ed., *Diary and Autobiography of John Adams,* IV:70–71.

13. Deas, ed., *Correspondence of Mr. Ralph Izard;* Hanger, "The Izards," 72–73; Wylma Wates, "Precursor to the Victorian Age: The Concept of Marriage and Family as Revealed in the Correspondence of the Izard Family of South Carolina," in Bleser, ed., *In Joy and Sorrow,* 4–14.

14. Cheves, "Izard of South Carolina," 214–17.

15. Charles Manigault, "Some Things Relating to my Family Affairs," folder 9, p. 15, Manigault Family Papers (SHC); Josephine du Pont to Margaret Manigault, Oct. 23, 1800, in Low, "Of Muslins and Merveilleuses," 69.

16. Margaret Manigault to Josephine du Pont, Apr. 13, 1800, in Low, "Of Muslins and Merveilleuses," 57. Manigault, "The Manigault Family of South Carolina from 1685 to 1886," 81–83.

17. Ralph Izard to John Vaughan, Aug. 25, 1793, Vaughan-Madeira Collection, John Vaughan Papers, box 1 (APS); Gabriel Manigault travel journal, [Oct.] 21, 1789, Manigault Family Papers (SCL); Charles Manigault, "Some Things Relating to our Family Affairs," folder 9, p. 17, Manigault Family Papers (SHC).

18. Charles Manigault, "Some Things Relating to our Family Affairs," folder 9, p. 18, Manigault Family Papers (SHC); "Reminiscences of Joshua Francis Fisher," box 10, J. Francis Fisher Section, Cadwalader Collection (HSP).

19. John Vaughan to Gabriel Manigault, Sept. 1, 1807, and Alice Izard to Margaret Manigault, Nov. 11, 1807, Manigault Family Papers (SCL); in the 1808 directory Gabriel Manigault's listing was 90 S. Eighth Street; in 1810 "Mrs. Gabriel Manigault, Gentlewoman," was listed on 453 High (Market) Street; in that year "Mrs. Izard, Gentlewoman," lived adjacently at 453 High Street; living in South Carolina until 1816, she had no separate listing until 1818.

20. Charles Manigault, "Some Things Relating to our Family Affairs," folder 9, p. 14, Manigault Family Papers (SHC); Wates, "Precursor to the Victorian Age," 6–8.

21. "Reminiscences of Joshua Francis Fisher," box 10, J. Francis Fisher Section, Cadwalader Collection (HSP).; Margaret Manigault to Henry Manigault, Mar. 16, 1816, Louis Manigault Papers (Duke).

22. Gabriel Manigault to Gabriel Henry Manigault, Dec. 17, 1808, Louis Manigault Papers (Duke).

23. With considerable financial success for himself but disastrous results for his slaves; Clifton, ed., *Life and Labor on Argyle Island;* Dusinberre, *Them Dark Days;* Young, "Ideology and Death on a Savannah River Rice Plantation"; Charles Manigault, "Some Things Relating to our Family Affairs," folder 9, p. 18, Manigault Family Papers (SHC).

24. Margaret Manigault to Charles Manigault, Dec. 1, 1811, Manigault Family Papers (SCL); Charles Manigault, "Souvenirs of our Ancestors, & of my immediate family," folder 9, pp. 11, 14, Manigault Family Papers (SHC).

25. Margaret Manigault to Charles Manigault, Jan. 14, 1812, Manigault Family Papers (SCL).

26. Charles Manigault, "Souvenirs of our Ancestors & of my immediate family," folder 9, p. 5, Manigault Family Papers (SHC); Gabriel Manigault to Gabriel Henry Manigault, Dec. 8, 1808, Louis Manigault Papers (Duke); Harriet Manigault Diary, Aug. 13 and Oct. 30, 1814 (HSP).

27. Margaret Manigault to Gabriel Henry Manigault, Apr. 9, 1816, Louis Manigault Papers (Duke).

28. Alice Izard to Margaret Manigault, Mar. 23, 1808, Mar. 31, 1811, Manigault Family Papers (SCL); Schloesser, *The Fair Sex;* Nash, "Rethinking Republican Motherhood."

29. Mary Huger Middleton to Margaret Manigault, n.d., Fisher Section, Cadwalader Collection (HSP).

30. Alice Izard to Margaret Manigault, Jan. 15, 1809 (Vaughan); July 29, 1809 (Biddle), Manigault Family Papers (SCL).

31. Alice Izard to Margaret Manigault, July 29, 1810 (T. L. Smith); July 10, 1810 (Lespinasse), Manigault Family Papers (SCL). Izard was referring to Julie de Lespinasse, *Lettres de Mademoiselle de Lespinasse.*

32. Todd, *Physical Culture and the Body Beautiful.*

33. Alice Izard to Margaret Manigault, Dec. 25, 1808 ("bucking"); Jan. 2, 1809 (Bernard) (SCL).

34. Alice Izard to Margaret Manigault, Mar. 10, 1811, Manigault Family Papers (SCL); on the intellectual qualities of this circle, see O'Brien, *Conjectures of Order,* I:97–100.

35. Perhaps Leprince du Beaumont, *Education Complette, ou Abrégé de l'Histoire Ancienne.*

36. The account in this paragraph is taken from the Harriet Manigault Diary, July 29, 1814 ("I will now give an exact account of how I spend my time") (HSP); Wakefield, A*n Introduction to Botany.*

37. Chateaubriand, *Travels in Greece, Palestine, Egypt, and Barbary;* Barthélemy, *Travels of Anacharsis the Younger.*

38. Hamilton, *Memoirs of Modern Philosophers.*

39. *Letters of Anna Seward.*

40. Margaret Manigault to Elizabeth Morris, Oct. 28, 1812, Manigault, Morris, and Grimball Family Papers (SHC); also titled *Select Reviews of Literature* or simply *Select Reviews,* the magazine was published between 1809 and 1812; it was followed by the *Analectic Magazine.*

41. Margaret Manigault to Josephine du Pont, Feb. 8, 1800, in Low, "Of Muslins and Merveilleuses," 54; Alice Izard to Margaret Manigault, Dec. 25, 1808, Manigault Family Papers (SCL); in late 1808 the girls were "to have gone to the Roman Catholic chapel, to hear the music there, under the guardianship of Mr. J[oseph] A[llen] Smith, but the rain has prevented them this Sunday, as it did the last" (Alice Izard to Margaret Manigault, Dec. 10, 1808, ibid.).

42. Margaret Manigault to Charlotte Georgina Smith, Nov. 20, 1813, Manigault Family Papers (SCL).

43. Alice Izard to Margaret Manigault, Jan. 9, 1809, Manigault Family Papers (SCL); Margaret Manigault to Alice Izard, Feb. 9, 1805, Feb. 13, 1805 (reference to George), Ralph Izard Family Papers (LC).

44. Hansen, A *Very Social Time,* explores these phenomena from the perspective of ordinary New Englanders.

45. Alice Izard to Margaret Manigault, Oct. 26, 1815; Aug. 20, 1805, Manigault Family Papers (SCL); Gabriel Manigault to Gabriel Henry Manigault, Dec. 9, 1808, Louis Manigault Papers (Duke).

46. Harriet Manigault to Elizabeth Morris, Aug. 4, 1815, Manigault Family Papers (SCL).

47. Eleanor Parke Custis to Elizabeth Bordley, July 2, 1797, and Custis to Bordley, n.d., Eleanor Parke (Custis) Lewis Papers (HSP).

48. Charlotte and Margaret Manigault to Elizabeth Morris, Apr. 11, 1816, Manigault Family Papers (SCL); Shields, *Civil Tongues and Polite Letters.*

49. Adams, ed., *Life and Writings of Jared Sparks,* 134; "Reminiscences of Joshua Francis Fisher," box 10, J. Francis Fisher Section, Cadwalader Collection (HSP).

50. Shackelford, *Jefferson's Adoptive Son;* Davis, *The Abbé Corrêa in America.*

51. Margaret Manigault to Elizabeth Morris, Mar. 24, 1813, Manigault, Morris, and Grimball Family Papers (SHC); Margaret Manigault to Elizabeth Morris, Nov. 7, 1812, and Margaret Manigault to Alice Izard, July 12, 1812, Manigault Family Papers (SCL).

52. Margaret Manigault to Josephine du Pont, Jan. 27, 1814 (Vaughan), June 19, 1814 (Elba), in Low, "The Youth of 1812," 199, 201; Margaret Manigault to Elizabeth Morris, Nov. 7, 1812, and Margaret Manigault to Alice Izard, July 12, 1812 (last quotation), Manigault Family Papers (SCL); Margaret Manigault to Elizabeth Morris, Mar. 24, 1813, ("moralizes") Manigault, Morris, and Grimball Family Papers (SHC).

53. Margaret Manigault to Josephine du Pont, Apr. 20, 1814, in Low, "Youth of 1812," 199.

54. Dowling, *Literary Federalism in the Age of Jefferson.*

55. Margaret Manigault to Josephine du Pont, June 19, 1814, in Low, "The Youth of 1812," 198, 193.

56. Schafly, "The First Russian Diplomat in America"; del Rio, *La Mision de Don Luis de Onis.*

57. Margaret Manigault to Charlotte-Georgina Smith, Jan. 23, 1814 ("astonishment"), Nov. 20, 1814, Manigault Family Papers (SCL).

58. Margaret Manigault to Charlotte-Georgina Smith, Dec. 6, 1813 (SCL), and Margaret Manigault to Josephine du Pont, Mar. 21, 1813, in Low, "Youth of 1812," 193.

59. Hunt, ed., *The First Forty Years of Washington Society,* 140.

60. Margaret Manigault to Charlotte-Georgina Smith, Sept. 11, 1814, Manigault Family Papers (SCL).

61. Margaret Manigault to Charlotte-Georgina Smith, Jan. 23, 1814, Manigault Family Papers (SCL); Pinckney, ed., *The Letterbook of Eliza Lucas Pinckney,* 80.

62. Margaret Manigault to Charlotte-Georgina Smith, Jan. 23, 1814, and Sept. 25, 1814, Manigault Family Papers (SCL); Charles Manigault, "Souvenirs of our Ancestors," folder 9, p. 3, Manigault Family Papers (SHC).

63. "Reminiscences of Joshua Francis Fisher," box 10, J. Francis Fisher Section, Cadwalader Collection (HSP); Alice Izard to Margaret Manigault, Jan. 13, 1808 ("French manners"), and Margaret Manigault to Elizabeth Morris, Mar. 26, 1821 (Mrs. Derby),

Manigault Family Papers (SCL); Charles Manigault, "Souvenirs of our Ancestors," folder 9, p. 12, Manigault Family Papers (SHC); on this point more generally, see Shields, *Civil Tongues and Polite Letters.*

64. Alice Izard to Margaret Manigault, Mar. 31, 1811, Manigault Family Papers (SCL).

65. Harriet Manigault to Elizabeth Morris, Aug. 10, 1814, Manigault, Morris, and Grimball Family Papers (SHC); Josephine du Pont to Manigault, Sept. 10, 1800, in Low, "Of Muslins and Merveilleuses," 66–67.

66. Harriet Manigault Diary, Dec. 20, 1813 (guests), Dec. 22, 1813 ("Tea Fight"), Jan. 12, 1814, Feb. 2, 1814 (partners); Charlotte Manigault to Elizabeth Morris, Apr. 11, 1816 (fainting); Alice Izard to Margaret Manigault, n.d. ("protector"); Margaret Manigault to Charlotte-Georgina Smith, Dec. 19, 1812 ("great deal of wine"), Nov. 20, 1813 ("sweet meats"), Feb. 16, 1821 ("rarity"), all in Manigault Family Papers (SCL).

67. Margaret Manigault to Charlotte-Georgina Smith, Aug. 11, 1816, and Harriet Manigault to Elizabeth Morris, Jan. 24, 1813, Manigault Family Papers (SCL); the Manigaults became intimate with the Biddles through the Craig family, whose daughter married Nicholas Biddle and through whom he acquired Andalusia; in 1813 Margaret characterized Nicholas Biddle as "an important young man"; Manigault to Josephine du Pont, Mar. 21, 1813, in Low, "Youth of 1812," 193.

68. [Margaret Manigault to Elizabeth Morris], Dec. 21, 1813, Manigault, Morris, and Grimball Family Papers (SHC); Alice Izard to Margaret Manigault, Dec. 7, 1808, Manigault Family Papers (SCL).

69. Alice Izard to Margaret Manigault, Mar. 24,1811, and Harriet & Charlotte-Georgina Manigault to Elizabeth Morris, Jan. 24, 1813, Manigault Family Papers (SCL).

Chapter 2: The Fisher Family Circle

1. Martineau, *Society in America,* I:172–73.

2. Bushman, *Refinement of America;* Hemphill, *Bowing to Necessities;* Miller, "An 'Uncommon Tranquility of Mind.'" Philip Dormer Stanhope, Lord Chesterfield (1694–1773), wrote letters of advice to his illegitimate son, which were published in 1774. Focusing on gentility rather than virtuous behavior, the letters were an immediate sensation on both sides of the Atlantic. Eighteen editions were published in America before 1800.

3. Wainwright, ed., *A Philadelphia Perspective,* 259 (entry for Aug. 15, 1856).

4. Wainwright, ed., A *Philadelphia Perspective,* 150 (entry for Dec. 29, 1843); Wainwright, "Sidney George Fisher: The Personality of a Diarist"; Doerflinger, A *Vigorous Spirit of Enterprise.*

5. J. Francis Fisher to George Harrison, Sept. 14, 1838, Brinton Coxe Collection, box 18 (HSP). Cadwalader, ed., *Recollections of Joshua Francis Fisher;* Harrison, ed., *Best Companions,* 13–16.

6. Entries for Dec. 29, 1843 ("superior"), Nov. 2, 1836 (Butler), in Wainwright, ed.,

Philadelphia Perspective, 150, 10; J. Francis Fisher to Elizabeth Fisher, Nov. 5, 1830 ("richer classes"), J. Francis Fisher Section, Cadwalader Collection (HSP); Trapier, *Incidents in My Life,* 10 (Harvard); Sidney George Fisher to J. Francis Fisher, July 29, 1832, box 19, Brinton Coxe Collection (HSP); Meigs, *Life of Charles Jared Ingersoll,* 47–52.

7. Greenberg, "Aristocrat as Copperhead"; Stevens, "The Webster-Ingersoll Feud"; Miegs, *Life of Charles Jared Ingersoll.*

8. Wainwright, ed., *Philadelphia Perspective,* 59 (entry for Sept. 12, 1838); on the Middletons, see Harrison, ed., *Best Companions,* 2–13, Cheves, "Middleton of South Carolina," and Lane, "The Middletons of Eighteenth-Century South Carolina."

9. Williams Middleton to J. Francis Fisher, Mar. 25, 1848, box 19, Brinton Coxe Collection (HSP).

10. Middleton et al., *Life in Carolina and New England during the Nineteenth Century.*

11. Shaffer, *To Be an American;* Murrin, "A Roof Without Walls."

12. Carson, "Early American Tourists and the Commercialization of Leisure," in Carson et al., eds., *Of Consuming Interests,* 373–405.

13. J. Francis Fisher to Elizabeth P. Fisher, Nov. 5, 1830, box 1, folder 1, Fisher Section, Cadwalader Collection (HSP); on religion and class in the South, see Ownby, *Subduing Satan,* Lewis, *The Pursuit of Happiness,* and Heyrman, *Southern Cross;* on the North, see Lazerow, "Rethinking Religion and the Working Class," and Zuckerman, "Holy Wars, Civil Wars."

14. Child, *The American Frugal Housewife,* 89; J. Francis Fisher to Elizabeth P. Fisher, Nov. 5, 1830, box 1, folder 1, Fisher Section, Cadwalader Collection (HSP); Mary H. Middleton to Septima Rutledge, Apr. 7, 1822, box 5, folder 9, ibid; Wainwright, ed., *Philadelphia Perspective,* p. 44 (entry for Feb. 4, 1838).

15. Trapier, *Incidents in My Life,* p. 10. Glover, "An Education in Southern Masculinity"; Scarborough, *Masters of the Big House.*

16. Mary H. Middleton to Elizabeth and Catherine Middleton, Feb. 20, 1822, Hering Family Papers (SCHS); Wainwright, ed., *Philadelphia* Perspective, 59 (entry for Sept. 12, 838); J. Francis Fisher to George and Sophia Harrison, Mar. 6, 1839, box 1, Fisher Section, Cadwalader Collection (HSP); on "select companies," see Bushman, "American High Style and Vernacular Cultures."

17. Sidney George Fisher to J. Francis Fisher, July 29, 1832, Brinton Coxe Collection (HSP).

18. Henry Middleton to Eliza Middleton, Nov. 10, 1836, box 7, Fisher Section, Cadwalader Collection (HSP); J. Francis Fisher to Sophia Harrison, Nov. 13–16, 1830, Dec. 12–17, 1830, box 1, ibid; Arthur married Paulina Bentivoglio; Harry wed Ellen Goggin of England; and Edward married Edwardina de Normann; for biographical information see Harrison, ed., *Best Companions,* xxxix–xli, and Cheves, "Middleton of South Carolina," 246–51.

19. Williams Middleton to Joshua Francis Fisher, Mar. 25, 1848, box 19, Brinton Coxe Collection (HSP) (first and last quotations); Williams Middleton to Eliza M. Fisher, Jan. 30, 1840, Hering Family Papers (SCHS).

20. Blumin, *Emergence of the Middle Class,* 234–38; Jaher, *Urban Establishment;* Baltzell, *Philadelphia Gentlemen.*

21. Wainwright, ed., *Philadelphia Perspective,* 65 (entry for Dec. 13, 1838); Eliza M. Fisher to Mary H. Middleton, Jan. 23, 1840, in Harrison, ed., *Best Companions,* 104; Bell, *Major Butler's Legacy,* 42–81.

22. "Reminiscences of Joshua Francis Fisher," box 10, J. Francis Fisher Section, Cadwalader Collection (HSP) (first quotation); A Member of the Philadelphia Bar, *Wealth and Biography of the Wealthy Citizens of Philadelphia,* 6 (second); A Merchant of Philadelphia, *Memoirs and Auto-Biography of some of the Wealthy Citizens of Philadelphia,* 11; Wainwright, ed., *Philadelphia Perspective,* 86–87 (entry for Oct. 24, 1839); Bell, *Major Butler's Legacy,* 227–54.

23. On Butler and the Kembles, see Bell, *Major Butler's Legacy;* Clinton, *Fanny Kemble's Civil Wars;* Blainey, *Fanny & Adelaide.*

24. Butler, *Journal,* II:136; Butler, *Journal of a Residence on a Georgian Plantation;* Clinton, ed., *Fanny Kemble's Journals,* 162–89; Bell, *Major Butler's Legacy,* 288–310.

25. Rebecca Gratz to Miriam Gratz Moses Cohen, Sept. 13, 1845, folder 4, Miriam Gratz Moses Cohen Papers (SHC); Charlotte Manigault Wilcocks's Diary, Sept. 10, 1842 (HSP).

26. Wainwright, ed., *Philadelphia Perspective,* 168, 86 (entries for May 15, 1844, Oct. 24, 1839); Eliza M. Fisher to Mary H. Middleton, Feb. 10, 1845, in Harrison, ed., *Best Companions,* 427.

27. Eliza M. Fisher to Mary H. Middleton, Jan. 23, 1840, Jan. 3, 1840, in Harrison, ed., *Best Companions,* 104, 95; Wainwright, ed., *Philadelphia Perspective,* 186, 113 (entries for Jan. 18, 1846, Jan. 26, 1841).

28. Emily Heyward Drayton Taylor, *The Draytons of South Carolina and Philadelphia.*

29. Joseph Wharton to Anna Lovering, Feb. 13, 1853, in Ingle, ed., "Joseph Wharton Goes South," 327; Wharton, *Genealogy of the Wharton Family of Philadelphia.*

30. Emily W. Sinkler to Thomas I. Wharton, Mar. 15, 1843, in LeClercq, ed., *Between North and South,* 33.

31. Wainwright, ed., *Philadelphia Perspective,* 27 (Robinson), 87 (Legaré), 187 (Carter's Party) (entries for Mar. 16, 1837; Oct. 16, 1839; Jan. 20, 1846).

32. Wainwright, ed., *Philadelphia Perspective,* 127 (entry for Nov. 14, 1841); Boykin, ed., *Victoria, Albert, and Mrs. Stevenson.*

33. Wainwright, ed., *Philadelphia Perspective,* 203 (entry for Dec. 12, 1847); Eliza M. Fisher to Marh H. Middleton, Feb. 1, 1842, in Harrison, ed., *Best Companions,* 244; Misses Middleton to Phoebe Rush, n.d., Misses Izard to Phoebe Rush, n.d., and "List of Names of persons in Philadelphia kept by Mrs. Rush for the convenience of inviting to parties between the years 1830 & 1840," Phoebe Rush Papers (LCP); Maria Davis's father was José Vidal, a Louisiana planter; I would like to thank Kathie Blankenstein of Natchez, Mississippi, who kindly supplied information about the Davises in a personal communication of Feb. 10, 1994.

34. Wyatt-Brown, *Southern Honor;* Hackney, "Southern Violence"; Ayers, *Vengeance and Justice.*

35. Wells, *The Origins of the Southern Middle Class,* 80–86; Cramer, *Concealed Weapon Laws of the Early Republic;* Williams, *Dueling in the Old South,* chapter 5.

36. Grayson quoted in Wyatt-Brown, *Southern Honor,* 351; Olsen, *Political Culture and Secession in Mississippi;* Baptist, *Creating an Old South.*

37. Rebecca Gratz to Ann Boswell Gratz and Benjamin Gratz, June 18, [1849], in Philipson, ed., *Letters of Rebecca Gratz,* 360–61.

38. Harriet, "On a Young Gentleman who was Killed in a Duel," *Philadelphia Album, and Ladies' Weekly Gazette* 1, no. 9 (1826): 8; A Lady (Charleston, S.C.), "On Dueling," ibid., 1, no. 22 (1826): 2–3.

39. Webb, *History of Pennsylvania Hall,* 142 (first quotation); Wainwright, ed., *Philadelphia Perspective,* 49–50 (entry for May 19, 1838); Richards, *"Gentlemen of Property and Standing,"* and Grimsted, *American Mobbing,* trace the public reaction to anti-abolitionist mobs in the antebellum North.

40. William H. Keating to Joel R. Poinsett, Apr. 1, 1830, Joel Roberts Poinsett Papers (HSP); A Merchant of Philadelphia, *Wealth and Biography of Wealthy Citizens of Philadelphia,* 6.

41. Wainwright, ed., *Philadelphia Perspective,* 223, 294–95 (entries for Apr. 17, 1849, Mar. 6, 1858); Eliza M. Fisher to Williams Middleton, Feb. 7, 1851, box 5, folder 4, Middleton Place Collection; J. Francis Fisher to Williams Middleton, Jan. 9, 1856, Hering Family Papers (SCHS) ("doubtful reputation").

42. Joshua Francis Fisher to Williams Middleton, Feb. 7, 1851; Eliza M. Fisher to Williams Middleton, Feb. 11, 1851, box 5, folder 4, Middleton Place Collection; Wainwright, ed., *Philadelphia Perspective,* 230 (entry for Apr. 17, 1849).

43. Wainwright, ed., *Philadelphia Perspective,* 230 (entry for Apr. 17, 1849); Arthur Middleton to Williams Middleton, Feb. 11, 1851, box 5, folder 4, Middleton Place Collection.

44. Corner, ed., *The Autobiography of Benjamin Rush,* 225; on southern gentility, see Lewis, *Ladies and Gentlemen on Display;* Grant, *North over South,* and Bushman, *Refinement of America,* 390–98, assess northern prejudices against southern gentility.

45. J. Francis Fisher to George Harrison, Sept. 21, 1838, box 18, Coxe Collection (HSP); Wainwright, ed., *Philadelphia Perspective,* 26 (Robinson), 48–49 (Masons) (entries for Mar. 16, 1837, Apr. 12, 1838).

46. "Reminiscences of Joshua Francis Fisher," box 10, J. Francis Fisher Section, Cadwalader Collection (HSP); Wainwright, ed., *Philadelphia Perspective,* 10, 278, 310, 317 (entries for Nov. 2, 1836, Sept. 19, 1857, Dec. 6, 1858, Feb. 17, 1859).

47. Joshua Francis Fisher Diary, Jan. 20, 1848 (HSP); Wainwright, ed., *Philadelphia Perspective,* 204 (entry for Feb. 13, 1848).

48. "Reminiscences of Joshua Francis Fisher," box 10, J. Francis Fisher Section, Cadwalader Collection (HSP); Wainwright, ed., *Philadelphia Perspective,* 47 (entry for Mar. 13, 1838).

49. Wainwright, ed., *Philadelphia Perspective,* 122–23 (entry for Aug. 8, 1841); Hebe, the daughter of Zeus and Hera, was worshiped as the goddess of youth.

50. "Boston, the Literary Emporium," *American Ladies' Magazine and Literary Gazette* 6 (1833): 140; Charlotte Manigault Wilcocks's Diary, Oct. 13, 1842 (HSP); Wainwright, ed., *Philadelphia Perspective,* 49 (entry for Apr. 28, 1838).

51. "Reminiscences of Joshua Francis Fisher," box 10, J. Francis Fisher Section, Cadwalader Collection (HSP); Wainwright, ed., *Philadelphia Perspective,* 81, 225 (Cadwalader), 255 (Olmsted), 211 (Baltimore) (entries for Apr. 17, 1839, Aug. 12, 1849, Mar. 17, 1856, June 17, 1848).

52. Henry Middleton Jr. to Eliza Middleton, June 27, 1836, box 1, Fisher Section, Cadwalader Collection (HSP); Williams Middleton to J. Francis Fisher, Aug. 3, 1851, Middleton Place Collection; Wainwright, ed., *Philadelphia Perspective,* 229 (entry for Dec. 16, 1849).

53. Eliza M. Fisher to Mary H. Middleton, Dec. 15 [1841–42], Hering Family Papers (SCHS) ; Eliza M. Fisher to Mary H. Middleton, Dec. 19, 1844, in Harrison, ed., *Best Companions,* 416.

54. Franklin, A *Southern Odyssey;* McCardell, *The Idea of a Southern Nation.*

55. Wainwright, ed., *Philadelphia Perspective,* 102 (entry for June 28, 1840).

56. Mary H. Middleton to Eliza M. Fisher, Apr. 21, 1844, in Harrison, ed., *Best Companions,* 379; Williams Middleton to J. Francis Fisher, June 30, 1851, Middleton Place Collection; J. Francis Fisher to John Brown Francis, Dec. 12, 1860, box 18, Coxe Collection (HSP).

57. Wainwright, ed., *Philadelphia Perspective,* 257, 260 (entries for May 27, 1856 [Sumner], June 7, 1856 [Kansas], Sept. 17, 1856 [secession]).

58. Ibid., 262 (entries for Oct. 8 and Oct. 18, 1856); Williams Middleton to J. Francis Fisher, Aug. 3, 1851, folder 5, box 5, and Fisher to Middleton, Mar. 1, 1855, folder 2, box 6, Middleton Place Collection.

Chapter 3: "Your appropriate sphere as a lady"

1. Alice Izard to Margaret Manigault, Mar. 23, 1808, Manigault Family Papers (SCL).

2. Ainsley Hall to Langdon Cheves, Mar. 10, 1819, Langdon Cheves Papers (SCHS).

3. Benjamin Rush, "Thoughts upon Female Education, Accommodated to the Present State of Society, Manners, and Government in the United States of America," in Rudolph, ed., *Essays on Education in the Early Republic,* 28.

4. Kerber, *Women of the Republic,* 185–231; Nash, "Rethinking Republican Motherhood."

5. Kaestle, *Pillars of the Republic;* Katznelson and Weir, *Schooling for All.*

6. Mott, *Observations on the Importance of Female Education,* 15; on women's education, see Bunkle, "Sentimental Womanhood and Domestic Education," Ryan, *Cradle of the Middle Class,* chapter 4, and Scott, "The Ever Widening Circle"; for a contrary

view, see Hendrick, "Ever-Widening Circle or Mask of Oppression?"; on the South, see Farnham, *The Education of the Southern Belle,* Wells, *The Origins of the Southern Middle Class,* chapter 6, and Clinton, "Equally Their Due."

7. "Essay on American Language and Literature," *North American Review* 1, no. 3 (Sept. 1815): 308.

8. Accounts of Eliza Ann Spragins, 1839–1842, Spragins Family Papers (VHS); Mrs. A. Sigoigne Account Books, 1837–39, p. 6 (DCL); Bowie, "Madame Grelaud's French School"; Elijah Smith to Nathaniel Ware, Oct. 29, 1820, Samuel Brown Papers (FHL); Margo and Villaflor, "The Growth of Wages in Antebellum America," 880; Margo, *Wages and Labor Markets in the United States.*

9. Hall, *Travels in North America,* II:340.

10. Phelps, *The Female Student,* 366; "Boston, the Literary Emporium," 140.

11. Thomas G. Percy to John William Walker, May 29, 1821, John William Walker Papers (ADAH); Israel Pickens to General Lenoir, May 31, 1826, quoted in Clinton, "Equally Their Due," 59; Stowe, *Intimacy and Power in the Old South,* 132.

12. Bowie, "Madame Grelaud's French School"; Rosengarten, *French Colonists and Exiles in the United States;* Hebert, "The French Element in Pennsylvania"; on French schools in Baltimore, see Hartridge, "St. Domingan Refugees in Maryland."

13. Emma Shannon to Levina Morris Shannon, Oct. 21, 1857, box 3, folder 33, Crutcher-Shannon Family Papers (MDAH). Johnson, "Madame Rivardi's Seminary in the Gothic Mansion," 10 (quotation).

14. King, *Lily,* 69.

15. Samuel Brown to Margaretta Brown, Feb. 22, 1813, June 23, 1813, Samuel Brown Papers (FHL).

16. Martha Richardson to James P. Screven, July 22, 1821, Aug. 7, 1822, Arnold and Screven Family Papers (SHC).

17. Clinton, "Equally Their Due," 59; Harriet Horry Ravenel quoted in Steven Stowe, "City, Country, and Feminine Voice," 295; John Williams Walker to Samuel Brown, Apr. 19, 1819, Samuel Brown Papers (FHL).

18. Rebecca Gratz to Maria Gist Gratz, July 27, 1841, in Philipson, ed., *Letters of Rebecca Gratz,* 241; William Gaston to Susan Gaston, Dec. 9, 1822, William Gaston Papers (SHC); Appie [Virginia] King to Anna King, Mar. 21, 1852, Thomas Butler King Papers (SHC).

19. Isabella Mease to James Iredell, May 24, 1830, and James Mease to John Iredell, May 20, 1830, James Iredell Sr. and Jr. Papers (Duke).

20. William Gaston to Joseph Hopkinson, Aug. 26, 1823, Hopkinson Family Papers (HSP).

21. Eleanor Parke Lewis to Elizabeth Bordley, Jan. 3, 1815, Eleanor Parke (Custis) Lewis Letters (HSP); on young Eleanor Lewis's death, see Lewis to Elizabeth Bordley Gibson, Nov. 22, 1820, ibid.

22. Virginia Shelby to Susan Shelby, Nov. 23, 1845, Grigsby Family Papers (FHL); Mary H. Gill to Mr. and Mrs. Wiley, Nov. 17, 1853, Charles J. Harris Papers (Duke).

23. William Stuckey to Marmaduke Shannon, Dec. 14, 1857, box 3, folder 34, Crutcher-Shannon Family Papers (MDAH).

24. "Address, to the Graduating Class, at St. Mary's Hall, Sept., 1849," in *St. Mary's Hall, Twenty-Sixth Term,* 33.

25. Steven M. Stowe, "The Not-So-Cloistered Academy," 94–95.

26. See the admonitions against "enthusiasm" among adolescent girls in Child, *The Mother's Book,* 130–61, a widely read antebellum advice book (quotation on 130).

27. Mary H. Gill to Mrs. and Mr. Wiley, Nov. 17, 1853, Harris Papers (Duke); Emma Shannon to Levina Morris Shannon, Oct. 21, Nov. 7, 1857, box 3, folder 33, Crutcher-Shannon Family Papers (MDAH).

28. Emma Shannon to Levina Morris Shannon, Oct. 21, 1857, box 3, folder 3, and Oct. 29 [1858], box 4 folder 45, Crutcher-Shannon Family Papers (MDAH); for evidence of the Shannons' friendship with northerners, see Mollie Starke to Anna Shannon, Oct. 19, 1858, box 4, folder 44, ibid.; Rebecca Gratz to Maria Gist Gratz, July 27, 1841, in Philipson, ed., *Letters of Rebecca Gratz,* 291.

29. Susan Polk to Sarah Polk, folder 25, Polk and Yeatman Family Papers (SHC); Jennie Ellis to Powhatan Ellis, Jan. 2, 1837, Jane Shelton Ellis Tucker Papers (VHS); Margaret Mordecai to Ellen Mordecai, Sept. 29, 1837, Margaret Mordecai Devereux Papers (SHC).

30. William Gaston to Susan Gaston, Dec. 9, 1822, William Gaston Papers (SHC); Elizabeth Buchanan to Hester Van Bibber, Aug. 27, 1816, Hester E. (Van Bibber) Tabb Papers (Duke).

31. Smith-Rosenberg, "The Female World of Love and Ritual."

32. Sally W. Clay to Elizabeth Spragins, Sept. 22, 1842, Spragins Family Papers (VHS).

33. Fern, *Fresh Leaves,* 69; Bushman, *The Refinement of America,* 300.

34. Joseph Hopkinson to Emily Hopkinson, Dec. 28, 1815, Hopkinson Family Papers (HSP); Mary H. Gill to Mary Wiley, June 21, 1855, Harris Papers (Duke); *Young Ladies Seminary.*

35. Richard D. Arnold to Ellen Arnold, Nov. 5, 1849, in Shryock, ed., *Letters of Richard D. Arnold,* 34–35; Staughton, *An Address Delivered . . . at Mrs. Rivardi's Seminary,* 8.

36. Doane, *Appeal to Parents for Female Education,* 23; William Polk to Mary Polk, July 1, 1823, Polk, Badger, and McGehee Family Papers (SHC); Mary H. Gill to Mary Wiley, June 21, 1855, Harris Papers (Duke); full publication information on the works listed by Gill may be found in the bibliography.

37. James Brooks, "Conversation," *Ladies' Companion* 8 (Jan. 1838): 123; Julia Marie Watson to Eliza Riddle, June 5, 1839, section ten, Claiborne Family Papers (VHS).

38. Gaston probably refers to Englishman Jeremiah Joyce's *Scientific Dialogues,* a youth-oriented text that first appeared in an American edition in 1815.

39. Jennie Ellis to Powhatan Ellis, Jan. 2, 1837, Jane Shelton Ellis Tucker Papers (VHS); Susan Gaston to William Gaston, Dec. 28, 1822, William Gaston Papers (SHC); Sallie B. to Mary Wiley, Feb. 3, 1855, Harris Papers (Duke).

40. Rosalie S. Calvert to Jean Charles Steir, Feb. 24, 1813, Dec. 16, 1815, in Hoyt, ed., "The Calvert-Steir Correspondence," 269–70; John Williams Walker to Samuel Brown, Apr. 19, 1919, Samuel Brown Papers (FHL).

41. Rush, "Thoughts upon Female Education," in Rudolph, ed., *Essays on Education,* 35; Jonathan Stevens, "A Letter from New England, Part I," *Ladies' Repository* 12 (Jan. 1852): 23; "The American Ideal Woman," *Putnam's Monthly Magazine of American Literature, Science, and Art* 2 (Nov. 1853): 530; on American attitudes to France, see Jones, *America and French Culture,* Chew, "Life in France between 1780 and 1815," Elkins and McKitrick, *The Age of Federalism,* 303–73, and Hale, "Many Who Wandered in Darkness."

42. "The American Ideal Woman," 530; John E. Warren to "My Dear Friend," May 27, 1850," in Warren, "Rambles in the Peninsula, No. III," *International Magazine of Literature, Art, and Science* 1 (July 29, 1850), 136; Maria Walker to Matilda Pope Walker, Feb. 8, 1824, Sept. 11, 1824, John Williams Walker Family Papers (ADAH).

43. For example, Tristrim L. Skinner to Joseph Skinner, Dec. 5, 1835, folder 16, Skinner Family Papers (SHC).

44. Jennie Ellis to Powhatan Ellis, Jan. 2, 1837, Jane Shelton Ellis Tucker Papers (VHS); Isabella Mease to James Iredell Jr., May 24, 1830, James Iredell Sr. and Jr. Papers (Duke); Susan Gaston to William Gaston, Dec. 28, 1822, William Gaston Papers (SHC).

45. Phelps, *The Female Student,* 366.

46. James Louis Petigru to Jane Petigru North, May 21, 1839, in Carson, ed., *Life, Letters, and Speeches of James Louis Petigru,* 204; see also Pease and Pease, *James Louis Petigru,* 75; on the Draytons, see O'Brien, "Politics, Romanticism, and Hugh Legaré," in his *Rethinking the South,* 63.

47. Y. E. Transou to Julie Conrad, Oct. 25, 1848, folder six, Jones Family Papers (SHC).

48. Alice Izard to Margaret Manigault, Aug. 26, 1805, Manigault Family Papers (SCL); William Polk to Mary Polk, Apr. 10, 1824, and William Polk to Sarah Polk, Mar. 6, 1822, Polk, Badger, and McGehee Family Papers (SHC).

49. Particularly in Foster, *The Coquette,* though this character remained popular throughout the antebellum period; see Mrs. H. C. Gardner, "The Ill-Bred Girl," *Ladies' Repository* 15 (Apr. 1855): 205–6; Smith-Rosenberg, "Domesticating Virtue."

50. Joseph Hopkinson to William Gaston, Sept. 2, 1823, William Gaston Papers (SHC); Gaston to Hopkinson, June 4, 1823, Hopkinson Family Papers (HSP).

51. Elizabeth Buchanan to Hester Van Bibber, Apr. 18, 1817, Tabb Papers; King, *Lily,* 73 ("Walk like a lady of good breeding, and pay no attention to the loafers"; the French has been transcribed accurately, including the grammatical and spelling errors); on this point, see Kasson, *Rudeness and Civility,* 93.

52. Bushman, *Refinement of America,* 359; Susan Polk to Sarah Polk, June 18, 1835, folder 25, Polk and Yeatman Family Papers (SHC).

53. Emma Shannon to Levina Morris Shannon, Apr. 8, 1858, box 4, folder 40, Crutcher-Shannon Family Papers (MDAH); Susan Polk to Sarah Polk, Apr. 4, 1835, Polk and Yeatman Family Papers (SHC); Accounts of Eliza Ann Spragins, 1839–42, Spragins Family Papers (VHS); Sidney Gill to Dr. I.B. Wiley, June 23, 1854, Harris Papers (Duke).

54. Georgia Bryan to Thomas Marsh Forman, June 8, 1823, folder 37, Arnold and Screven Family Papers (SHC); Charles Picot to James Iredell, Aug. 15, 1832, James Iredell Sr. and Jr. Papers (Duke); Tristrim Lowther Skinner to Joseph Blount Skinner, Dec. 5, 1835, folder 16, Skinner Family Papers (SHC).

55. Eliza M. Fisher to Mary H. Middleton, Jan. 23, 1840, Apr. 12, 1842, in Harrison, ed., *Best Companions,* 105, 262.

56. When word reached Virginia of Nicholas Biddle's death in 1843, a schoolmate of Eliza Spragins wrote that "no one will lament his death more than" the Sigoignes (Mattie [?] to Eliza Spragins, Mar. 14, 1843, Spragins Family Papers [VHS]).

57. Wyatt-Brown, *The House of Percy,* 98.

58. [?] to Mary Wiley, Feb. 11, 1855, Harris Papers (Duke).

59. Richard D. Arnold to Ellen Arnold, Jan. 11, 1850, in Shryock, ed., *Letters of Richard D. Arnold,* 37.

Chapter 4: The Republic of Medicine

1. Wertenbaker, *Princeton.*

2. On medicine, see Warner, *The Therapeutic Perspective,* and Kett, *The Formation of the American Medical Profession;* on the law, see Cook, *The American Codification Movement;* on the ministry, see Scott, *From Office to Profession,* and Hatch, *The Democratization of American Christianity.*

3. Ezell, "A Southern Education for Southrons"; Warner, "A Southern Medical Reform"; Duffy, "Sectional Conflict and Education in Louisiana"; Duffy, "A Note on Ante-Bellum Southern Nationalism"; Breeden, "States-Rights Medicine in the Old South."

4. Benjamin Rush to Dr. James Currie, Apr. 27, 1800, Joseph Lyon Miller Papers (VHS).

5. Norwood, *Medical Education in the United States;* Ludmerer, *Learning to Heal;* Pease and Pease, "Intellectual Life in the 1830s," in O'Brien and Moltke-Hansen, eds., *Intellectual Life in Antebellum Charleston,* 243–49.

6. Warner, "Orthodoxy and Otherness"; Haller, *Medical Protestants;* Rosner, "Thistle on the Delaware."

7. Jones, "American Doctors and the Parisian Medical World, 1830–1840," 46; Warner, *Against the Spirit of the System,* 38–39; Bonner, *American Doctors and German Universities.*

8. *New Orleans Medical News and Hospital Gazette,* 3 (1856–57); 678, cited in Warner, "A Southern Medical Reform," 214; on the argument for environmental medical specificity, see Warner, "The Idea of Southern Medical Distinctiveness."

9. "The New Orleans School of Medicine," *New Orleans Medical News and Hospital Gazette* 3 (1856–57), cited in Warner, "A Southern Medical Reform," 214; John Dabney to Jane Dabney, Feb. 23, 1818, Charles William Dabney Papers (SHC); "Medical Statistics of the State of Virginia," *Southern Literary Messenger* 8 (Oct. 1842): 643–44; on sectionalism in other educational fields, see McCardell, *The Idea of a Southern Nation,* chapter 5.

10. Daniel Drake to Samuel Brown, Nov. 3, 1818, Samuel Brown Papers (FHL).

11. Richard D. Arnold to Dr. D. F. Condie, Apr. 10, 1847, in Shryock, ed., *Letters of Richard D. Arnold,* 31; Arthur H. Shaffer, "David Ramsay and the Limits of Revolutionary Nationalism," in O'Brien and Moltke-Hansen, eds., *Intellectual Life in Antebellum Charleston,* 54.

12. Ludmerer, *Learning to Heal;* Norwood, *Medical Education;* Corner, *Two Centuries of Medicine;* Barlow and Powell, "A Dedicated Medical Student."

13. Neil McNair to Annabella McNair, Nov. 4, 1838, Annabella McNair Papers (Duke); Robert Nelson to Eliza Nelson, Dec. 17, 1843, Elizabeth K. Nelson Papers (VHS); Charles Bonner to John Young Bassett, Oct. 21, 1842, John Young Bassett Papers (SHC).

14. J.N. to N. McClelland, Feb. 26, 1830, McClelland Family Papers (SHC) (first quotation); Josiah Nott to James Gage, Mar. [1837], folder 3, James McKibbin Gage Papers (SHC); Charles Bonner to John Young Bassett, Oct. 21, 1842, John Young Bassett Papers (SHC); on the clinical opportunities offered by schools in southern towns, see Warner, "A Southern Medical Reform," 217–20.

15. Warren, A *Doctor's Experiences in Three Continents,* 129 (first quotation); W. B. Miller to Charles Earl Johnson, Dec. 30, 1849, Charles Earl Johnson Papers (Duke); Clarence Robards to Clarence Lytton Robards & Alice Tannehill, Oct. 13, 1850, Daniel S. Hill Papers (last quotation) (Duke); on student misbehavior at antebellum colleges, see Pace, *Halls of Honor,* chapters 3 and 4, and Drinkwater, "Honor and Student Misconduct in Southern Antebellum Colleges."

16. James C. Billingslea, "An Appeal on Behalf of Southern Medical Colleges and Southern Medical Literature," *Southern Medical and Surgical Journal* new series 12 (Sept. 1856): 398; Harvey L. Byrd, "Patronage of Northern Medical Schools by Southern Students of Medicine," *Oglethorpe Medical and Surgical Journal* 2 (Nov. 1859): 235, cited in Breeden, "Rehearsal for Secession," 176; "Home Education at the South," *De Bow's Review* 10 (Mar. 1851): 362; Leland, *Memoirs,* 136.

17. R. N. Venable to "Dear Father and Mother," Feb. 19, n.d., Carrington Family Papers (VHS).

18. David Hamilton to Sarah Hamilton, Dec. 31, 1837, folder 3, Benjamin C. Yancey Papers (SHC).

19. David Hamilton to Sarah Hamilton, May 23, 1838, folder 3, Yancey Papers (SHC) (first quotation); Marmaduke Kimbaugh to Nathaniel Hunt, Dec. 17, 1858, Nathaniel Hunt Papers, Duke University (second quotation); on the destruction of Pennsylvania Hall; see Webb, *History of Pennsylvania Hall,* and Bacon, "Lucretia

Mott," 63–79; on antiabolition mobs, see Richards, *"Gentlemen of Property and Standing."*

20. Carrie Fries to Francis Fries, [Mar.–Apr. 1860], Fries-Shaffner Family Papers (SHC); Harriet and Georgina Manigault to Elizabeth Morris, Jan. 24, 1813, Manigault Family Papers (SCL).

21. Francis T. Grady to John L. Powell, n.d., John L. Powell Papers (VHS); on the large proportion of southern students, see (in addition to the above two letters) James Herbert Gregory to Francis R. Gregory, Nov. 12, 1825, Ferebee, Gregory, and McPherson Papers (SHC), and David Hamilton to Sarah Hamilton, Dec. 31, 1837, Yancey Papers (SHC).

22. J. Quarles to Prof. [L. S.] Joynes, H[unter] H. McGuire and F[rancis] E. L[uckett] to Dr. David H. Tucker, and McGuire and Luckett to Tucker, all Dec. 17, 1859, Minutes of the Board of Visitors to the Medical College of Virginia (MCV); on the incident, see Breeden, "Rehearsal for Secession?" and "States-Rights Medicine in the Old South," Duffy, "Sectional Conflict and Medical Education in Louisiana," 299–300, and Duffy, "A Note on Ante-Bellum Southern Nationalism and Medical Practice," 273.

23. Thomas F. Lee, Chairman et al., and Drs. Luckett and McGuire, to Governor Henry Wise, Dec. 19, 1859, Minutes of the Board of Visitors to the Medical College or Virginia (MCV); Breeden, "Rehearsal for Secession?" 180–86; Matriculation Records of the Medical College of Virginia, 6th Session, 1859–60 (MCV). Forty-four percent of the seceding students who enrolled at the Medical College of Virginia were native Virginians. Other states represented were Alabama (17), North Carolina (16), Mississippi (16), South Carolina (14), Georgia (6), Arkansas (4), and Tennessee, Louisiana, Missouri, and Texas (2 each); on the figures from Jefferson, see *Catalogue of the Trustees, Professors, and Students* (1860).

24. Breeden, "Rehearsal for Secession?" 201; *North American and United States Gazette* (Philadelphia), Dec. 22, 1850, and *Baltimore American,* Dec. 26, 1859; on the sums disbursed to the students, see the Sanger Historical Files, Secession of Philadelphia Students (MCV).

25. *Baltimore American,* Dec. 30, 1859; "Southern Medical Students in Northern Medical Colleges—Once More," *Oglethorpe Medical and Surgical Journal* 3 (Jan. 1861): 268–69. The enrollment figures were culled from *Catalogue of the Trustees, Professors, and Students of the Jefferson Medical College of Philadelphia* (1861), and *Catalogue of the Trustees, Officers, and Students of the University of Pennsylvania* (1861).

26. Wyatt-Brown, *The House of Percy,* 10.

27. Theophilus E. Beesley et al. (forty-four signatories) "To the Honorable, the Board of Trustees of the University of Pennsylvania," Aug. 3, 1818, folder 1500, box 18, UPA 3 (Penn); on Hare's social attitudes, see his *Defence of the American Character,* and Cox, "Vox Populi," 259–72.

28. Alexander Edmiston to Margared M. Edminston, Dec. 18, 1808, Alexander Edmiston Papers (FHL).

29. Richard D. Arnold to Alfred Stillé, Mar. 4, 1860, in Shryock, ed., *Letters of Richard D. Arnold,* 93; William Browne to Dr. Grattan (?), Dec. 9, 1818, section 8, Claiborne Family Papers (VHS); William Bankhead et al. (26 signatories) to Thomas T. Hewson, Feb. 26, 1819, folder 1501, UPA 3: General Administration (Penn).

30. Fabius Haywood to [?], Apr. 10, 1825, folder 113, Ernest Haywood Collection of Haywood Family Papers (SHC).

31. Wyatt-Brown, *Southern Honor,* 294; on class, regional, and gendered aspects of illicit behavior by men in this period, see Stowe, *Intimacy and Power in the Old South,* 82, Miller, "'An Uncommon Tranquility of Mind,'" Hemphill, "Middle Class Rising," Walters, "The Erotic South," Gorn, "'Good-Bye, Boys, I Die a True American," and Cohen, *The Murder of Helen Jewett.*

32. J.N. to Nathaniel McClelland, Feb. 26, 1830, McClelland Family Papers (SHC); James Madison Brannock to George Brannock, Jan. 26, 1851, James Madison Brannock Papers (VHS).

33. B. T. Gunter to Edward Barksdale, Oct. 17, 1848, Peter Barksdale Papers (Duke); Stephen Davis to Dr. John Owen, Jan. 28, 1810, Campbell Family Papers (Duke); Edward Pegrame to Francis Gregory, Sept. 7, 1827, Ferebee-Gregory-McPherson Papers (SHC); on upper-class expectations of entitlement on this score, focusing on the early republic, see Stansell, *City of Women,* 20–30.

34. Josiah Nott to James Gage, Mar. [1837], folder 3, James McKibbin Gage Papers (SHC).

35. Robert Nelson to Eliza Nelson, Dec. 17 [1843–44], Elizabeth K. Nelson Papers (VHS); W. B. Miller to Charles Earl Johnson, Nov. 16, 1849, Charles Earl Johnson Papers (Duke) ("*rich* sect" and knife fights); Thomas Cash to his brother, Dec. 14, 1853, Jarratt-Puryear Family Papers (Duke); Robert E. Peyton to Dr. F. Peyton, Jan. 31, 1827, Peyton Family Papers (VHS).

36. Elias Boudinot Stockton et al. to the Trustees of the University of Pennsylvania (a petition of 100 medical students asking that James Rush, M.D., fill the chair of theory and practice of medicine vacated by Benjamin Smith Barton's death) Dec. 1, 1815, folder 1499, box 18, UPA 3 (Penn); on the "informal curriculum," see Farnham, *Education of the Southern Belle,* 120–45.

37. James M. Brannock to James Brannock Sr., Nov. 17, 1850, James Madison Brannock Papers (VHS); on honor and hospitality, see Wyatt-Brown, *Southern Honor,* 331–39.

38. Fabius Julius Haywood to Elizabeth Haywood, Feb. 28, 1825, folder 112, and William Haywood Jr. to Fabius Julius Haywood, Apr. 5, 1827, folder 125, Haywood Family Papers (SHC).

39. William Haywood Jr. to Fabius Julius Haywood, Apr. 5, 1825, folder 125, Haywood Family Papers (SHC); Robert L. Coleman to Mary Eliza Fleming, Mar. 9, 1853, Mary Eliza (Fleming) Schooler Papers (Duke); invitation cards of Dr. [Benjamin Franklin] Bache, Jan. 5, 1852, and Dr. and Mrs. James Rush, Jan. 13, 1853, Daniel William Lassiter, Francis Rives Lassiter, and Charles Trotter Lassiter Papers (Duke);

Abner Grigsby to Lucian Grigsby, Dec. 23, 1844, Grigsby Family Papers (VHS); John Powell to Henry Curtis, Nov. 26, 1827, Henry Curtis Papers (VHS).

40. Clarence Robards to cousin, Nov. 11, 1850, Daniel S. Hill Papers (Duke); Abner Grigsby to Lucian Grigsby, Dec. 23, 1844, Grigsby Family Papers (VHS).

41. John D. Owen to Mrs. Mary Jane Owen, Jan. 20, 1847, Campbell Family Papers (Duke); Abner Grigsby to Lucian Grigsby, Dec. 23, 1844, Grigsby Family Papers (VHS). Rotundo, *American Manhood,* 201–202, and Rorabaugh, *The Alcoholic Republic,* discuss alcohol consumption and masculine sociability.

42. Benjamin Rush to David Ramsay, Nov. 5, 1778, in Brunhouse, ed., *David Ramsay, 1749–1815,* 56–57; Richard D. Arnold to Mrs. [?] Arnold, May 13, 1846, in Shryock, ed., *Letters of Richard D. Arnold,* 28–30.

43. Jordan, "University of Pennsylvania Men who Served in the Civil War."

44. Coleman Rogers to Samuel Brown, Nov. 3, 1818, Samuel Brown Papers (FHL); Richard D. Arnold to Ellen (Arnold) Cosens, Sept. 25, 1860, and Arnold to Gr. George T. Elliot Jr., Sept. 28, 1865, in Shryock, ed., *Letters of Richard D. Arnold,* 98 and 128.

Chapter 5: Science and Sociability

1. Furness, A *Discourse,* 14; see Warren, *Joseph Leidy,* for a biographical account of a figure who successfully navigated this intellectual shift.

2. Furness, A *Discourse,* 14.

3. *Transactions of the American Philosophical Society, Held at Philadelphia, for Promoting Useful Knowledge* 1 (1769–71): preface, xiv, xv; Baatz, "Philadelphia Patronage," 112; on Philadelphia science, see Bell, "The Scientific Environment in Philadelphia," Chinard, "The American Philosophical Society and the World of Science," May, *The Enlightenment in America,* 197–22, and Gross, "The American Philosophical Society and the Growth of Science."

4. Story, "Class and Culture in Boston."

5. Benjamin Vaughan to John Vaughan, Aug. 12, 1784, box 1, Madeira-Vaughan Papers (APS).

6. "Extract from Mrs. Deborah Logan's biographical account of George Logan, box 2, Madeira-Vaughan Papers (APS); on the Vaughans, see Craig C. Murray, *Benjamin Vaughan,* Geffen, *Philadelphia Unitarianism,* and Stetson, "The Philadelphia Sojourn of Samuel Vaughan."

7. Adams, ed., *Life and Writings of Jared Sparks,* 133–34; Caleb Forshey to John Vaughan, June 6, 1841, American Philosophical Society Archives (APS).

8. "Reminiscences of Joshua Francis Fisher," box 10, J. Francis Fisher Section, Cadwalader Collection (HSP); Furness, A *Discourse,* 12–13; John Vaughan to Thomas Jefferson, Mar. 28, 1801, Thomas Jefferson Papers (LC).

9. Corrêa da Serra to John Vaughan, Dec. 17, 1815, in Davis, *Abbé Corrêa in America,* 160; Benjamin Story to Samuel Brown, care of John Vaughan, May 3, 1822, Samuel Brown Papers (FHL); Thomas Jefferson to John Vaughan, June 17, 1817, Apr. 8, 1818, box 2, John Vaughan Papers (APS.); William Dunbar to John Vaughan, Dec. 15, 1806,

in Rowland, ed., *Life, Letters, and Papers of William Dunbar,* 349; "Reminiscences of Joshua Francis Fisher," box 10, J. Francis Fisher Section, Cadwalader Collection (HSP).

10. "Reminiscences of Joshua Francis Fisher," box 10, J. Francis Fisher Section, Cadwalader Collection (HSP); John Izard Middleton to Vaughan, May 12, 1826, George Izard to Vaughan, June 8, 1818, Henry Middleton to Vaughan, Apr. 12, 1831, and Joel Poinsett to Vaughan, Apr. 20, 1827 (offering to send material), Mar. 5, 1829 (introducing Mr. Del Rio from Mexico), all in the American Philosophical Society Archives (APS); James Ferguson to John Vaughan, Apr. 6, 1817, John Vaughan Papers (APS).

11. Thomas Cooper to John Vaughan, June 28, 1821, Miscellaneous Manuscripts (APS); Charles Short to Vaughan, Oct. 25, 1830, American Philosophical Society Archives (APS); William Preston to Vaughan, July 12, n.d., Miscellaneous Manuscripts (APS).

12. James Ramsay to John Vaughan, Sept. 25, 1818, Samuel Brown to Vaughan, June 10, 1802, and William Johnson to Vaughan, Apr. 24, 1811, all in the American Philosophical Society Archives (APS).

13. Stephen Elliott to John Vaughan, n.d., American Philosophical Society Collection of Broadsides, no. 80 (APS); on the influence of the American Philosophical Society as a model, see Greene, *American Science in the Age of Jefferson.*

14. Stephen Elliott to Joel Poinsett, Jan. 9, 1822, Joel R. Poinsett Papers (HSP); Elliott to Poinsett, Nov. 10, 1823, Miscellaneous Manuscripts (APS); on the weakness of scientific societies in the South and West, see Ewan, "The Growth of Learned Societies," Stephens, "Literary and Philosophical Society," Stephens, *Science, Race, and Religion in the American South,* and Wells, *Origins of the Southern Middle Class,* chapter 4.

15. Gross, "The American Philosophical Society and the Growth of Science," 177–78, Charles Short to John Vaughan, Dec. 16, 1839, and L. Gay Lussac to Vaughan, Nov. 28, 1838, Miscellaneous Manuscripts (APS); on professionalization, see Reingold, "Dangerous Speculations," in Oleson and Brown, eds., *The Pursuit of Knowledge in the Early American Republic.*

16. "Extract of a Letter from Mr. Dunbar to Mr. Vaughan," [1810], American Philosophical Society Archives (APS); on Dunbar, see Rowland, ed., *Life, Letters, and Papers of William Dunbar* and DeRosier, "William Dunbar."

17. William Dunbar to John Vaughan, Oct. 20, 1805, Dunbar to Vaughan, n.d. [1806], and Dunbar to [?], Apr. 28, 1806, in Rowland, ed., *Life, Letters, and Papers of William Dunbar,* 327, 328–29, 340.

18. For vivid depictions of these areas, see Baptist, *Creating an Old South,* introduction and chapter 1, Olsen, *Political Culture and Secession in Mississippi,* chapter 1, and Morris, *Becoming Southern,* chapters 1 and 2.

19. John Quitman to John Vaughan, June 26, 1840, Miscellaneous Manuscripts (APS); Caleb Forshey to John Vaughan, June 6, 1841, American Philosophical Society Archives (APS); Harry Toumlin to John Vaughan, May 13, 1822, Miscellaneous Manuscripts (APS); on Forshey, see Gross, "The American Philosophical Society and the Growth of Science," 177–78.

20. [?] to George Izard & family, Apr. 24, 1807, Journal of the Proceedings of the Corresponding Secretaries of the American Philosophical Society, American Philosophical Society Archives, I:93 (APS); George Izard to Joseph Hopkinson, Nov. 28, 1827, Hopkinson Family Papers (HSP).

21. Corrêa da Serra to Joel R. Poinsett, Oct. 20, 1819; *Charleston Mercury,* July 18, 1827; Thomas Cooper to John Vaughan, Oct. 24, 1832; Joel R. Poinsett to John Vaughan, Aug. 2, 1833; Thomas Cooper to John Vaughan, Feb. 16, 1836 (all letters in Miscellaneous Manuscripts, APS).

22. John Vaughan to Wade Hampton, Sept. 13, 1832, American Philosophical Society Archives (APS).

23. *Transactions of the American Philosophical Society* 1 (1769–71): preface, iii, xvii; Peter DuPonceau and Clement Biddle et al., to the Membership Committee of the American Philosophical Society, July 18, 1823, Letters of Nomination for Membership, American Philosophical Society Archives (APS); Peter DuPonceau to Joel R. Poinsett, Nov. 9, 1826, Poinsett Papers (HSP); DuPonceau and Nathaniel Chapman to [Membership Committee], Dec. 15, 1820, and Joseph Hopkinson et al. to [Membership Committee], n.d., both in Letters of Nomination for Membership, American Philosophical Society Archives (APS).

24. William Tilghman, "Report of the Historical and Literary Committee to the American Philosophical Society," *Transactions of the American Philosophical Society* new series 1 (1818): xi–xii; Stephen DuPonceau to Thomas Jefferson, Nov. 14, 1815, Historical and Literary Committee Letterbooks, American Philosophical Society Archives, I:1 (APS).

25. Entries for July 14, 1815, and May 8, 1816, I:5, 12, 36, Minutes of the Historical and Literary Committee, American Philosophical Society Archives (APS); Stephen DuPonceau to General David B. Mitchell, Oct. 12, 1818 (regarding Eliza Tunstall), and DuPonceau to Tunstall, Jan. 11, 1819, Historical and Literary Committee Letterbooks, II:20–21, II:22–23, American Philosophical Society Archives (APS); entry for Feb. 2, 1817, Minutes of the Historical and Literary Committee, II:3, American Philosophical Society Archives (APS).

26. W. B. Bullock to General David B. Mitchell, Oct. 10, 1818, Historical and Literary Committee Letterbooks, II:23, American Philosophical Society Archives (APS); Tilghman, "Report of the Historical and Literary Committee," xiii, American Philosophical Society Archives (APS).

27. "Reminiscences of Joshua Francis Fisher," box 10, J. Francis Fisher Section, Cadwalader Collection (HSP); Adams, ed., *Life and Writings of Jared Sparks,* 133–34; William Dunbar to John Vaughan, June 26, 1840, Miscellaneous Manuscripts (APS).

28. Tilghman, A*n Eulogium, in Commemoration of Doctor Caspar Wistar,* 35; Tyson, *Sketch of the Wistar Party,* 6; on Wistar, see also Hosack, *Tribute to the Memory of the Late Caspar Wistar,* and Caldwell, A*n Eulogium on Caspar Wistar.*

29. Tyson, *Sketch of the Wistar Party,* 6–7.

30. Ibid., 8–9, 10–11, 24.

31. Posner, ed.,"Philadelphia in 1830," 242–43; Hall, *Travels in North America,* II:339; Hamilton, *Men and Manners in America,* I:342–43; Daubeny, *Journal of a Tour through the United States,* 90–91.

32. Tilghman, *An Eulogium in Commemoration of Doctor Caspar Wistar,* 44. Robert Walsh to John Vaughan, n.d. [Friday morning], Miscellaneous Manuscripts (APS); on Furness, see Geffen, *Philadelphia Unitarianism,* 186–91; on similar attitudes toward race and slavery in New England, see Melish, *Disowning Slavery.*

33. Tyson, *Sketch of the Wistar Party,* lists the association's members by year of induction.

34. "List of Guests Invited to Wistar Party, Oct. 1848," Miscellaneous Manuscripts of the Wistar Party (APS); Thomas Percy to John Williams Walker, June 12, 1821, John Williams Walker Papers (ADAH); William Dillingham, "The Life of Judge Gaston," Memoirs of Deceased Members (APS).

35. Gross, "The American Philosophical Society and the Growth of Science," 304–305.

36. *Chronicle of the Union League of Philadelphia,* 38; Henry Copee to Isaac Lea, Sept. 25, 1862, George Sharswood to Lea, Apr. 24, 1863, and Moncure Robinson to Lea, Sept. 26, 1861, all in the Manuscript Archives of the Wistar Association, American Philosophical Society Archives (APS).

37. Hugh B. Grigsby to Charles Trego, May 1, 1857, Letters Acknowledging Election, American Philosophical Society Archives (APS).

Chapter 6: ". . . all the world is a city"

1. John Izard Middleton to Nathaniel Russell Middleton, May 26, June 8, 1834, Nathaniel Russell Middleton Papers (SHC).

2. Barbara Carson, "Early American Travelers and the Commercialization of Leisure," in Carson, Hoffman, and Alberts, ed., *Of Consuming Interests;* Chambers, *Taking the Waters;* Lewis, *Ladies and Gentlemen on Display;* Brown, *Inventing New England;* Aron, *Working at Play;* Sears, *Sacred Places.*

3. Eliza Haywood to John Haywood, June 11, 1824, Haywood Family Papers (SHC); Conway Robinson to James Alfred Jones, Sept. 11, 1853, Conway Robinson Letterbook, Robinson Family Papers, 1836–1899 (VHS); *A Guide to the Lions of Philadelphia,* v, 6.

4. See also Chambers, *Taking the Waters;* by contrast, sociability at mostly southern-patronized resorts such as the Virginia Springs reinforced sectional feelings; Lewis, *Ladies and Gentlemen on Display.*

5. William Gilmore Simms to George Frederick Holmes, Aug. 15, 1842, in Oliphant, Odell, and Eaves, eds., *Letters of William Gilmore Simms,* I:319; Wylie, ed., *Memoirs of Judge Richard H. Clark,* 94; I thank Charles J. Johnson for bringing this source to my attention.

6. David R. Goldfield, "Pursuing the American Dream: Cities in the Old South," in Brownell and Goldfield, eds., *The City in Southern History,* 52–91; Amos, *Cotton*

City; O'Brien and Moltke-Hansen, eds., *Intellectual Life in Antebellum Charleston;* James, *Antebellum Natchez;* on ambivalence to cities, see Steven Stowe, "City, Country, and the Feminine Voice," in O'Brien and Moltke-Hansen, eds., *Intellectual Life in Antebellum Charleston,* 195–304.

7. Hugh Merritt Rose to Henry Rose, Oct. 16, 1825, Hugh Merritt Rose Papers (VHS); Shryock, ed., *Letters of Richard D. Arnold,* 9.

8. Nicholas B. Wainwright, "The Age of Nicholas Biddle, 1825–1841," and Elizabeth M. Geffen, "Industrial Development and Social Crisis, 1841–1854," both in Weigley, ed., *Philadelphia.*

9. Drury Lacy to Williana Lacy, May 16, 1839, Drury Lacy Papers (SHC); Elizabeth Ruffin Diary, Aug. 1, 1827, in O'Brien, ed., *An Evening When Alone,* 76; James Henry Hammond to Catherine Hammond, Apr. 17, 1836, James Henry Hammond Letters (SHC); Matilda Hamilton Diary, Feb. 25, 1857, Hamilton Family Papers (VHS).

10. Sheriff, *The Artificial River,* 24–25; Siry, *DeWitt Clinton and the American Political Economy,* 5; Howe, *The Political Culture of the American Whigs,* 48.

11. Kasson, *Rudeness and Civility,* 127–30, and Stansell, *City of Women,* 61–62, describe working-class behavior toward middling and upper-class pedestrians.

12. Myers, *Sketches on a Tour,* 442–43.

13. McKelvy, *American Prisons;* Lewis, *From Newgate to Dannemora;* Rothman, *The Discovery of the Asylum;* Hirsch, *The Rise of the Penitentiary.*

14. Myers, *Sketches on a Tour,* 443. Thibaut, "'To Pave the Way to Penitence'"; Kashatus, "'Punishment, Penitence and Reform"; Patrick, "Ann Hinson."

15. Matilda Hamilton Diary, Feb. 26, 1857, Hamilton Family Papers (VHS); John Strobia Diary, Sept. 17, 1817 (VHS).

16. Henry Massie Travel Journal, n.d. [1808] (VHS); J.Q.P. of N.C., "Extracts from Gleanings on the Way," *Southern Literary Messenger* 4 (Apr. 1838): 250; Waln, *The Hermit in America,* 78–79.

17. Anne J. Willing to Mary Byrd, Mar. 19, 1808, in Meade, ed., "The Papers of Richard Evelyn Byrd," 117; Harriet Manigault Diary, July 31, 1814 (HSP); Jane Caroline North Diary, Aug. 18, 1850, in O'Brien, *An Evening When Alone,* 194.

18. On this point, see Bushman, *The Refinement of America,* 352–55.

19. Dr. Adam Alexander Travel Diary, Oct. 6, 1801, Alexander-Hillhouse Papers (SHC); Redlich, "The Philadelphia Water Works"; Scharf and Westcott, *History of Philadelphia,* I:530; Seelye, *Beautiful Machine,* 7–8; *Lloyd's Mercantile Port Folio,* 7.

20. Myers, *Sketches on a Tour,* 446; Jane Caroline North diary, Aug. 20, 1850, in O'Brien, *An Evening When Alone,* 95; Albert G. Jefress to Augustus D. Clark, June 23, 1838, Clark Family Papers (VHS).

21. Fern, *Fresh Leaves,* 240.

22. C.G.P., A *Traveler's Sketch,* 13; John Strobia Diary, Sept. 11, 1817 (VHS).

23. Dr. Adam Alexander Travel Diary, Oct. 6, 1801, Alexander-Hillhouse Papers (SHC); Peale opened his museum in 1786 and administered it for his remaining forty-

one years in several locations, including the upper floor of Independence Hall; see Sellers, *Mr. Peale's Museum,* for a full description.

24. Moltke-Hansen, "The Expansion of Intellectual Life," in O'Brien and Moltke-Hansen, eds., *Intellectual Life in Antebellum Charleston,* 32.

25. Halttunen, *Confidence Men and Painted Women;* Kasson, *Rudeness and Civility.*

26. *A Guide to the Stranger,* 9, 12, 14 (LCP).

27. Harvey Washington Walter Diary, Sept. 1, 1849, Harvey Washington Walter Papers, and Andrew Polk to Sarah Polk, Jan. 12, 1842, Polk-Yeatman Family Papers (both SHC); Stansell, *City of Women,* 20–30.

28. Moltke-Hansen, "The Expansion of Intellectual Life," 8; on the international context of the southern feelings of distinctiveness, see Rugemer, "The Southern Response to British Abolitionism."

29. Shaffer, "Arthur Ramsay and the Limits of Revolutionary Nationalism," in O'Brien and Molke-Hansen, eds., *Intellectual Life in Antebellum Charleston,* 54.

30. *A Traveler's Tour through the United States,* 5, 6, 45.

31. Myers, *Sketches on a Tour,* 436, 449–50.

32. Ibid., 68; Travers, *Celebrating the Fourth;* Waldstreicher, *In the Midst of Perpetual Fetes;* Newman, *Parades and the Politics of the Street;* Branson, *These Fiery Frenchified Dames.*

33. Davison, *Fashionable Tour,* 438–39; Mary Telfair to Mary Few, Nov. 12, n.d., William Few Collection (GDHA); contemporary works that stress the distinctiveness of ordinary southerners include Olmsted, *The Cotton Kingdom,* and Trollope, *Domestic Manners of the Americans;* historical works that emphasize their unique outlook include Thornton, *Politics and Power,* Ford, *The Origins of Southern Radicalism,* Baptist, *Creating an Old South,* and Harris, *Plain Folk and Gentry;* Anderson, *Imagined Communities,* assesses the imaginative nature of feelings of national unity.

34. John Strobia Diary, Sept. 6, 1817 (VHS); Hobsbawm and Ranger, *The Invention of Tradition;* on struggles over the meaning of the revolution, see Morrison, "American Reactions to European Revolutions" and Cohen, *The Revolutionary Histories.*

35. On this point Snay, *Gospel of Disunion* and "American Thought and Southern Distinctiveness," are useful. Emma Shannon to Levina Morris Shannon, Apr. 8, 1858, box 4, folder 40, Crutcher-Shannon Family Papers (MDAH); J.Q.P. from N.C., "Extracts from Gleanings on the Way," 250; Briton Basil Hall also expressed dismay at the appearance of Independence Hall, although he ascribed its poor condition to Americans' failure to respect anything "on account of its age, or, indeed, on any other account. Neither historical associations, nor high public services, nor talents, nor knowledge, claim any particular reverence from the busy generations of the present hour" (Hall, *Travels in North America,* II:375–76).

36. Waldstreicher, *In the Midst of Perpetual Fetes;* Travers, *Celebrating the Fourth;* Koschnik, "Political Conflict and Public Contest"; Morrison, "American Reaction to European Revolutions."

37. *A Guide to the Lions of Philadelphia,* 20; Matilda Hamilton Diary, Feb. 24, 1857, Hamilton Family Papers (VHS); Hugh Merritt Rose to Henry Rose, Oct. 16, 1825, Hugh Merritt Rose Papers (VHS).

38. Clement Comer Clay to Hugh Lawson Clay, July 22, 1850, Clement Comer Clay Papers (Duke); on Clay's sectionalism, see McCardell, *The Idea of a Southern Nation,* 114–15, 275.

39. Chambers, *Things as They Are in America,* 317; William Brisland to William Ferriday, July 20, 1839, box 1, folder 15, Brisland-Shields Family Papers (MDAH); Mary Telfair to Mary Few, July 14, n.d., item 223, William Few Collection (GDHA).

40. Davison, *The Traveller's Guide through the Middle and Northern States,* 68; *A Guide to the Lions of Philadelphia,* 43.

41. Eliza E. Haywood to John Haywood, June 11, 1824, Haywood Family Papers (SHC); J.Q.P. from N.C., "Extracts from Gleanings on the Way," 251.

42. John Houston Bills Diary, July 11, 12 1845 (SHC); Benjamin L. C. Wailes Diary, July 4, 1859 (MDAH); *Ladies' Vase: or, Polite Manual for Young Ladies,* 22; William Gaston to Joseph Hopkinson, June 4, 1823, William Gaston Papers (SHC).

43. J.Q.P. from N.C., "Extracts from Gleanings on the Way," 251; Preachy R. Grattan to Jane E. Grattan, Sept. 27, 1837, Preachy R. Grattan Papers (SHC); Henry Massie Travel Journal, 1808 (VHS).

Epilogue

1. For more detailed assessments of developments in Philadelphia during the secession crisis and Civil War, see Kilbride, "Philadelphia and the Southern Elite," chapter 8, Gallman, *Mastering Wartime,* and especially Dusinberre, *Civil War Issues in Philadelphia.*

2. *Douglass' Monthly,* Feb. 1862; Willson, *Sketches of the Higher Classes of Coloured Society,* 16; Brewster in *North American* (Philadelphia), Jan. 17, 1861.

3. See especially Melish, *Disowning Slavery,* chapter 6.

4. Susan I. Lesley to Joseph Lyman, Dec. 2, 1850, in Ames, ed., *Life and Letters of Peter and Susan Lesley,* I:380; McClure, *Old Time Notes of Pennsylvania,* I:467; on the effect of events such as the Sumner incident, see Gienapp, "The Crime Against Sumner."

5. Arthur Ritchie to Harriet Ritchie, Aug. 8, 1850, folder 11, Murdock and Wright Family Papers (SHC); Lewis Morris Grimball to Elizabeth Grimball, Nov. 27, 1860, Grimball Family Papers (SHC); William B. Reed to Robert Gourdin, Jan. 22, 1861, in Schankman, "William B. Reed and the Civil War," 460–61.

6. *Proceedings of the Great Union Meeting,* 6; *Great Union Meeting, Philadelphia, December 12, 1859,* 1, 22; on the partisan realignment among the upper class, see Weigley, "The Border City in Civil War," in Weigley, ed., *Philadelphia,* 370, and Wainwright, "The Loyal Opposition in Civil War Philadelphia," 295.

7. Reed, *Speech on the Presidential Question,* 11–12; Gienapp, "Nativism and the Creation of a Republican Majority"; Anbinder, *Nativism and Slavery,* 263; Myers, "The Rise of the Republican Party in Pennsylvania."

8. Scharf and Westcott, *History of Philadelphia,* I:740–53; A Merchant of Philadelphia, *The Ides of March,* 18–19, 24 (peace); *Palmetto Flag,* Mar. 30, 1861; "A Singular Meeting," *North American,* Jan. 17, 1861; similar sentiments were voiced in New York; see Foner, *Business and Slavery,* and Bernstein, *The New York City Draft Riots,* 125–92.

9. Ingersoll, *Secession: A Folly and a Crime,* 8, 29; Wister, ed., "Sarah Butler Wister's Civil War Diary," 273–75.

10. Wister, *The Philadelphia Club,* 40–41.

11. Williams Middleton to Joshua Francis Fisher, Apr. 30, 1861, box 6, folder 2, Middleton Place Collection; Fisher, *Concessions and Compromises,* 1–2, 7, 9–11.

12. Elizabeth M. Fisher to Joshua Francis Fisher, Apr. 4, 1861, box 1, Fisher Section, Cadwalader Collection (HSP); "Message to Congress in Special Session," in Basler, ed., *The Collected Works of Abraham Lincoln,* IV:438; Joshua Fisher to John Brown Francis, July 11, 1861, box 18, Coxe Collection (HSP); Williams Middleton to Elizabeth M. Fisher, May 12, 1865, box 7, Fisher Section, Cadwalader Collection (HSP); Joshua Fisher to John Brown Francis, Oct. 15, 1863, Coxe Collection (HSP).

13. George Fanhestock Diary, June 4, 1863 (HSP); Wainwright, ed., *Philadelphia Perspective,* 445 (entry for Jan. 3, 1863), 498 (entry for May 23, 1865).

14. Elizabeth M. Fisher to Henry Bentivoglio Van Ness Middleton, Aug. 10, 1864, box 6, folder 11, Middleton Place Collection; Williams Middleton to Elizabeth M. Fisher, Aug. 6, 1865, box 7, folder 3, Middleton Place Collection; Williams to Elizabeth, June 30, 1865, box 18, Coxe Collection (HSP); Roark, *Masters without Slaves,* and Wayne, *The Reshaping of Plantation Society,* discuss the changes confronted by planters such as Williams Middleton throughout the South during the Civil War and Reconstruction.

15. Joshua Francis Fisher to Williams Middleton, Jan. 14, 1868, box 9, folder 1, Middleton Place Collection; Faust, *Mothers of Invention,* 274.

16. Joshua Francis Fisher to Williams Middleton, Jan. 14, 1868, box 9, folder 1, Middleton Place Collection; Blight, "For Something beyond the Battlefield"; Fisher's "Reminiscences" were privately published as *Recollections of Joshua Francis Fisher Written in 1864;* as Fisher's letter to Williams indicates, however, they seem to have been written not in 1864 but intermittently during the mid-to-late 1860s.

17. Joshua Francis Fisher to Williams Middleton, Jan. 14, 1868, box 9, folder 1, Middleton Place Collection; "Reminiscences of Joshua Francis Fisher," box 10, J. Francis Fisher Section, Cadwalader Collection (HSP).

18. *Chronicle of the Union League of Philadelphia,* 38, 39, 44.

19. Ibid., 25, 52; Lathrop, *History of the Union League,* 31.

Bibliography

Manuscript Sources

Alabama Department of Archives and History, Montgomery
John Williams Walker Family Papers
Georgia Department of History and Archives, Atlanta
William Few Collection
Special Collections, Filson Historical Society, Louisville, Kentucky
Samuel Brown Papers
Alexander Edmiston Papers
Grigsby Family Papers
Mississippi Department of Archives and History, Jackson
Brisland-Shields Family Papers
Crutcher-Shannon Family Papers
Benjamin L.C. Wailes Diary
Rare Book, Manuscript, and Special Collections Library, Duke University, Durham, North Carolina
Peter Barksdale Papers
Campbell Family Papers
Clement Comer Clay Papers
Charles J. Harris Correspondence, 1850–1913
Daniel S. Hill Papers
Nathan Hunt Papers
James Iredell Sr. and Jr. Papers
Jarratt-Puryear Family Papers
Charles Earl Johnson Papers
Daniel William Lassiter, Francis Rives Lassiter, and Charles Trotter Lassiter Papers
Louis Manigault Papers
Annabella McNair Papers
Eliza K. Nelson Papers
Mary Eliza (Fleming) Schooler Papers
Hester E. (Van Bibber) Tabb Papers

Southern Historical Collection, Manuscript Division, Davis Library, University of North Carolina, Chapel Hill
Alexander and Hillhouse Family Papers (#11)
Arnold and Screven Family Papers (#3419)
John Young Bassett Papers (#1527)
John Houston Bills Diary (#2245)
Miriam Gratz Moses Cohen Papers (#2639)
Margaret Mordecai Devereux Papers (#2492-z)
Ferebee, Gregory, and McPherson Papers (#3374)
Fries-Shaffner Family Papers (#4046)
James McKibbin Gage Papers (#1812-z)
William Gaston Papers (#272)
Preachy R. Grattan Papers (#3594)
Grimball Family Papers (#980)
James Henry Hammond Letters (#305-z)
Ernest Haywood Collection of Haywood Family Papers (#1290)
Jones Family Papers (#2884)
T. Butler King Papers (#1252)
Drury Lacy Papers (#3641)
Manigault Family Papers (#484)
Manigault, Morris, and Grimball Family Papers (#976)
McClelland Family Papers (#3869)
Nathaniel Russell Middleton Papers (#507)
Murdock and Wright Family Papers (#532)
Polk, Badger, and McGehee Family Papers (#3979)
Polk and Yeatman Family Papers (#606)
Skinner Family Papers (#669)
Harvey Washington Walter Papers (#3399)
Benjamin C. Yancey Papers (#25)
Department of Special Collections, Dickinson College Library, Carlisle, Pennsylvania
Mrs. A. Sigoigne Account Books, 1837–39, 1851–55
Manuscripts Department, American Philosophical Society, Philadelphia
American Philosophical Society Archives
American Philosophical Society Collection of Broadsides
Historical and Literary Committee Letterbooks
Journal of the Proceedings of the Corresponding Secretaries of the American Philosophical Society
Letters Acknowledging Election
Letters of Nomination for Membership
Manuscript Archives of the Wistar Association
Minutes of the Historical and Literary Committee
Memoirs of Deceased Members

Miscellaneous Manuscripts Collection
Miscellaneous Manuscripts of the Wistar Party
John Vaughan Papers, Vaughan-Madeira Collection

Historical Society of Pennsylvania, Philadelphia
Cadwalader Collection, J. Francis Fisher Section
Brinton Coxe Collection
George Fanhestock Diary
Joshua Francis Fisher Diary
Hopkinson Family Papers
Eleanor Parke (Custis) Lewis Letters
Harriet Manigault Diary
Joel Roberts Poinsett Papers
Charlotte Wilcocks Diary

Library Company of Philadelphia
Phoebe Rush Papers

University of Pennsylvania Archives and Records Center, Philadelphia
General Administration, University of Pennsylvania Archives
List of Medical Graduates in the University of Pennsylvania from the year 1814 to 1850, c. 1814–c. 1850
Recapitulation of the Number of Graduates since the year 1802 [through 1819]

Middleton Place Foundation, South Carolina
Middleton Place Collection

South Carolina Historical Society, Charleston
Langdon Cheves Papers
Hering Family Papers
George Izard Autobiography

South Caroliniana Library, University of South Carolina, Columbia
Ralph Izard Papers
Manigault Family Papers

Special Collections and Archives, Tompkins-McCaw Library, Medical College of Virginia
Matriculation Records of the Medical College of Virginia, 6th Session, 1859–60
Minutes of the Board of Visitors to the Medical College of Virginia
Sanger Historical Files, Secession of Philadelphia Students

Virginia Historical Society, Richmond
James Madison Brannock Papers
Carrington Family Papers
Claiborne Family Papers, 1803–1954
Clark Family Papers
Henry Curtis Papers
Grigsby Family Papers
Hamilton Family Papers

Henry Massie Travel Journal
Joseph Lyon Miller Papers
Elizabeth K. Nelson Papers
Peyton Family Papers
John L. Powell Papers
Robinson Family Papers, 1836–1899
Hugh Merritt Rose Papers
Spragins Family Papers
John Strobia Diary
Jane Shelton Ellis Tucker Papers
Library of Congress, Washington, D.C.
Ralph Izard Family Papers, 1778–1826
Thomas Jefferson Papers

Newspapers

Baltimore American
Douglass' Monthly (Rochester, N.Y.)
North American (Philadelphia)
Palmetto Flag (Philadelphia)

Published Primary Sources

Adams, Herbert Baxter, ed. *Life and Writings of Jared Sparks, Comprising Selections from His Journals and Correspondence.* Boston: Houghton Mifflin, 1893.

Agassiz, Louis, and Augustus A. Gould. *Principles of Zoology: Touching the Structure, Development, Distribution and Natural Arrangement of the Races of Animals, Living and Extinct, with Numerous Illustrations.* Boston: Gould, Kendall & Lincoln, 1848.

"The American Ideal Woman." *Putnam's Monthly Magazine of American Literature, Science, and Art* 2 (Nov. 1853): 527–32.

Ames, Mary Lesley, ed. *Life and Letters of Peter and Susan Lesley.* 2 vols. New York: Putnam, 1909.

Barthélemy, J. J. *Travels of Anacharsis the Younger in Greece, during the Middle of the Fourth Century before the Christian Era. . . .* Translated by George Belmont. 7 vols. London: G. G. & J. Robinson, 1791.

Basler, Ray P., ed. *The Collected Works of Abraham Lincoln.* 9 vols. Washington, D.C.: Lincoln Sesquicentennial Commission, 1959.

Billingslea, James C. "An Appeal on Behalf of Southern Medical Colleges and Southern Medical Literatures." *Southern Medical and Surgical Journal* new series 12 (Sept. 1856): 398–402.

Boykin, Edward, ed. *Victoria, Albert, and Mrs. Stevenson.* New York: Rinehart, 1957.

Brooks, James. "Conversation." *Ladies' Companion* 8 (Jan. 1838): 123–25.

Butler, Frances Anne. *Journal.* 2 vols. London: John Murray, 1835.

———. *Journal of a Residence on a Georgian Plantation in 1838–1839.* New York: Harper, 1863.

Butterfield, L. H., ed. *Diary and Autobiography of John Adams.* 4 vols. Cambridge: Belknap Press of Harvard University Press, 1961.

C.G.P. *A Traveler's Sketch.* Philadelphia: McLaughlin Brothers, 1861.

Cadwalader, Sophia., ed. *Recollections of Joshua Francis Fisher Written in 1864.* Boston: Privately printed, 1929.

Caldwell, Charles. *An Eulogium on Caspar Wistar, M.D., Professor of Anatomy, Delivered by Appointment before the Members of the Philadelphia Medical Society.* Philadelphia: Thomas Dobson & Son, 1818.

Carson, James Petigru, ed. *Life, Letters, and Speeches of James Louis Petigru.* Washington, D.C.: Lowdermilk, 1920.

Catalogue of the Trustees, Officers, and Students of the University of Pennsylvania, Session 1860–61. Philadelphia: Collins, 1861.

Catalogue of the Trustees, Professors, and Students of the University of Pennsylvania, Session 1859–60. Philadelphia: Collins, 1860.

Chambers, William. *Things as They Are in America.* Philadelphia: Lippincott, Gambo, 1854.

Chastellux, Marquis de. *Travels in North America in the Years 1780, 1781, and 1782.* Edited by Howard C. Rice Jr. 2 vols. Chapel Hill: University of North Carolina Press, 1963.

Chateaubriand, François-René. *Travels in Greece, Palestine, Egypt, and Barbary; during the Years 1806 and 1807.* Translated by Frederic Choberl. Philadelphia: M. Thomas, 1813.

Child, Lydia Maria. *The American Frugal Housewife: Dedicated to those who are Not Ashamed of Economy.* 12th ed. Boston: Carter, Hendee, 1833.

———. *The Mother's Book.* 2nd ed. Boston: Carter & Hendee, 1831.

Chronicle of the Union League of Philadelphia 1862 to 1902. Philadelphia: Privately printed, 1902.

Clifton, James M. ed. *Life and Labor on Argyle Island: Letters and Documents of a Savannah River Rice Plantation, 1833–1867.* Savannah: Beehive Press, 1978.

Clinton, Catherine, ed. *Fanny Kemble's Journals.* Cambridge: Harvard University Press, 2000.

———. *Fanny Kemble's Civil Wars.* New York: Simon and Schuster, 2000.

Corner, George W., ed. *The Autobiography of Benjamin Rush: His 'Travels Through Life' Together with his Commonplace Book for 1789–1813.* Princeton: Princeton University Press, 1948.

Crouse, Maurice Alfred. "The Manigault Family of South Carolina, 1685–1783." Ph.D. dissertation, Northwestern University, 1964.

Daubeny, Charles. *Journal of a Tour through the United States.* Oxford: T. Combe, 1843.

Davison, G. M. *The Fashionable Tour; or, A Trip to the Springs, Niagara, Quebeck, and Boston, in the Summer of 1821.* Saratoga Springs: G. M. Davison, 1822.

———. *The Traveller's Guide through the Middle and Northern States and the Provinces of Canada.* 7th ed. Saratoga Springs: G. M. Davison and S. S. & W. Wood, 1837.

Deas, Anne Izard, ed. *Correspondence of Mr. Ralph Izard, of South Carolina, from the Year 1774 to 1804, with a Short Memoir.* New York: Charles S. Francis, 1844.

Dick, Thomas. *Celestial Scenery, or, the Wonders of the Planetary System Displayed: Illustrating the Perfections of Deity and the Plurality of Worlds.* Philadelphia: E. C. Biddle, 1838.

———. *The Sidereal Heavens, and other Subjects Connected with Astronomy, as Illustrative of the Character of the Deity and of Infinity of Worlds.* Philadelphia: E. C. Biddle, 1838.

Doane, George Washington. *An Appeal to Parents for Female Education on Christian Principles, with a Prospectus of St. Mary's Hall, Green Bank, Burlington, New Jersey.* Burlington, N.J.: J. L. Powell, Missionary Press, 1837.

"Essay on American Language and Literatures." *North American Review* 1 (Sept. 1815): 307–14.

"Extracts from the Letters of Correspondents." *Southern Literary Messenger* 1 (Feb. 1835): 322–23.

Fern, Fanny [Sarah Payson (Willis) Parton]. *Fresh Leaves.* New York: Mason Brothers, 1857.

Fisher, Joshua Francis. *Concessions and Compromises.* Philadelphia: C. Sherman & Son, 1860.

Foster, Hannah Webster. *The Coquette; or, the History of Eliza Wharton.* Boston: Printed by Samuel Etheridge, 1797.

Furness, William H. *A Discourse, Delivered on the Occasion of the Death of John Vaughan in the First Congregational Unitarian Church, Sunday, Jan. 16, 1842.* Philadelphia: J. Crissy, 1842.

Gardner, Mrs. H. C. "The Ill-Bred Girl." *Ladies' Repository* 15 (Apr. 1855): 205–6.

Great Union Meeting, Philadelphia, December 12, 1859. Philadelphia: Crissy & Markley, n.d.

Griswold, Rufus Wilmont. *The Republican Court; or, American Society in the Days of Washington.* New York: Appleton, 1856.

A Guide to the Lions of Philadelphia; Comprising a Description of the Places of Amusement, Exhibitions, Public Buildings, Public Squares, &c. in the City; and of the Places of Public Resort and Objects of Interest and Curiosity in the Environs. Designed as a Pocket Cicerone for Strangers. Philadelphia: Thomas T. Ash, 1837.

A Guide to the Stranger, or Pocket Companion for The Fancy, Containing A List of the Gay Houses and Ladies of Pleasure in the City of Brotherly Love and Sisterly Affection. Philadelphia, [1849?].

Hall, Basil. *Travels in North America in the Years 1827 and 1828.* 3 vols. Edinburgh: Cadell, 1829.

Hamilton, Elizabeth. *Memoirs of Modern Philosophers.* Edited by Claire Grogan. 1800; Peterborough, Ont.: Broadview Press, 2000.

Hamilton, Thomas. *Men and Manners in America.* 2 vols. Edinburgh: Blackwood, 1833.

Hare, Robert. *Defence of the American Character: Or an Essay on Wealth as an Object of Cupidity or the Means of Distinction in the United States.* Philadelphia, 1819.

Harriet. "On a Young Gentleman Who Was Killed in a Duel." *Philadelphia Album, and Ladies' Weekly Gazette* 1, no. 9 (1826): 8.

Harrison, Eliza Cope, ed. *Best Companions: Letters of Eliza Middleton Fisher and Her Mother, Mary Hering Middleton, from Charleston, Philadelphia, and Newport, 1839–1846.* Columbia: University of South Carolina Press, 2001.

"Home Education in the South." *Debow's Review* 16 (Mar. 1851): 362–63.

Hosack, David. *Tribute to the Memory of the Late Caspar Wistar, M.D.* New York: C. S. Van Winkle, 1818.

Hoyt, William D., ed. "The Calvert-Steir Correspondence." *Maryland Historical Magazine* 38 (Sept. 1943): 123–40.

Hunt, Galliard, ed. *The First Forty Years of Washington Society, Portrayed by the Family Letters of Mrs. Samuel Harrison Smith (Margaret Bayard) from the Collection of her Grandson, J. Henley Smith.* New York: Scribner, 1906.

Hunt, Robert. *The Poetry of Science, or, Studies of the Physical Phenomenon of Nature.* London: Reeve, Benham, & Reeve, 1848.

Ingersoll, Joseph Reed. *Secession: A Folly and a Crime.* Philadelphia: King & Baird, 1861.

Ingle, H. Larry. "Joseph Wharton Goes South." *South Carolina Historical Magazine* 96 (1995): 304–28.

J.Q.P. from N.C. "Extracts from Gleanings on the Way." *Southern Literary Messenger* 4 (Apr. 1838): 249–51.

Joyce, Jeremiah. *Scientific Dialogues, Intended for the Instruction and Entertainment of Young People: In Which the First Principals of Natural and Experimental Philosophy are Fully Explained.* London, 1809; Philadelphia: M. Carey, 1815.

King, Susan Petigru. *Lily: A Novel, and Gerald Gray's Wife.* Edited by Jane H. Pease and William H. Pease. Durham: Duke University Press, 1993.

Ladies' Vase: or, Polite Manual for Young Ladies. Original and Selected. By an American Lady. Lowell, Mass.: N. L. Dayton, 1843.

Lady (Charleston, S.C.), A. "On Dueling." *Philadelphia Album, and Ladies' Weekly Gazette* 1, no. 22 (1826): 2–3.

Lathrop, George Parsons. *History of the Union League of Philadelphia, from its Origin and Foundation to the Year 1882.* Philadelphia: Lippincott, 1884.

LeClercq, Anne Sinkler Whaley. *Between North and South: The Letters of Emily Wharton Sinkler, 1842–1865.* Columbia: University of South Carolina Press, 2001.

Leland, Charles Godfrey. *Memoirs.* New York: Appleton, 1893.

Leprince de Beaumont, Jeanne-Marie. *Education Complette, ou Abrégé de l'Histoire Ancienne avec des Notions Géographiques et Chronologiques.* Amsterdam: Delechaux, 1818.

Lloyd's Mercantile Port Folio and Business Man's Guide. Designed to be a Book of Reference, for Western & Southern Merchants Trading with Philadelphia. Philadelphia: W. A. Lloyd, 1855.

Low, Betty-Bright P. "Of Muslins and Merveilleuses: Excerpts from the Letters of Josephine du Pont and Margaret Manigault." *Winterthur Portfolio* 9 (1974): 29–75.

———. "The Youth of 1812: More Excerpts from the Letters of Josephine du Pont and Margaret Manigault." *Winterthur Portfolio* 11 (1976): 173–212.

Manigault, Gabriel E. "The Manigault Family of South Carolina from 1685 to 1886." *Transactions of the Huguenot Society of South Carolina* 4 (1897): 48–84.

Martineau Harriet. *Society in America.* 3 vols. London: Saunders & Otley, 1837.

McClure, A. K. *Old Time Notes of Pennsylvania.* 2 vols. Philadelphia: John C. Winston, 1905.

Meade, Everard Kidder." The Papers of Richard Evelyn Byrd, I, of Frederick County, Virginia." *Virginia Magazine of History and Biography* 54 (1946): 106–18.

"Medical Statistics of the State of Virginia." *Southern Literary Messenger* 8 (Oct. 1842): 643–47.

Meigs, William Montgomery. *The Life of Charles Jared Ingersoll.* Philadelphia: J. B. Lippincott, 1897.

A Member of the Philadelphia Bar. *Wealth and Biography of the Wealthy Citizens of Philadelphia, Containing an Alphabetical Arrangement of Persons Estimated to be Worth $50,000 and Upwards, with the Sums Appended to Each Name; Being Useful to Bankers, Merchants, and Others.* Philadelphia: G. B. Zieber, 1845.

A Merchant of Philadelphia. *The Ides of March: or, Abraham Lincoln as a Private Citizen. Being a Sequel to the End of the Irrepressible Conflict.* Philadelphia: King & Baird, 1861.

———. *Memoirs and Auto-Biography of some of the Wealthy Citizens of Philadelphia, with a Fair Estimate of their estates—Founded upon a Knowledge of Facts.* Philadelphia: Published by the Booksellers, 1846.

Middleton, Alicia Hopton, et al. *Life in Carolina and New England During the Nineteenth Century. As Illustrated by the Reminiscences and Letters of the Middleton Family of South Carolina and of the De Wolf family of Bristol Rhode Island.* Bristol, R.I.: Privately printed, 1929.

Mott, Abigail Field. *Observations on the Importance of Female Education, and Maternal Instruction, with the Beneficial Influence on Society, Designed to Be Used as a Class Book.* 2nd ed. New York: M. Day, 1827.

Myers, J. C. *Sketches on a Tour through the Northern and Eastern States, the Canadas, and Nova Scotia.* Harrisonburg, Va.: J. H. Wartmann & Brothers, 1849.

Nichol, J. P. *The Architecture of the Heavens.* London: J. W. Parker, 1850.

O'Brien, Michael, ed. *An Evening When Alone: Four Journals of Single Women in the South.* Charlottesville: University of Virginia Press for the Southern Texts Society, 1993.

Oliphant, Mary C. Simms, Alfred Taylor Odell, and T. C. Duncan Eaves, eds., *The Letters of William Gilmore Simms.* 5 vols. Columbia: University of South Carolina Press, 1952–82.

Olmsted, Frederick Law. *The Cotton Kingdom: A Traveller's Observations on Cotton and Slavery in the American Slave States. Based on three Former Volumes of Journeys and Investigations.* New York: Mason Brothers, 1861.

Phelps, Almire Hart Lincoln. *The Female Student; or, Lectures to Young Ladies on Female Education.* New York: Leavitt, Lord, 1836.

Philipson, David, ed. *Letters of Rebecca Gratz.* Philadelphia: Jewish Publication Society of America, 1929.

Pinckney, Elise, ed., *The Letterbook of Eliza Lucas Pinckney 1739–1762.* Chapel Hill: University of North Carolina Press, 1972.

Posner, Russell M., ed. "Philadelphia in 1830: An English View." *Pennsylvania Magazine of History and Biography* 95 (1971): 239–43.

Proceedings of the Great Union Meeting in Philadelphia, on the 21st of November, 1850. Philadelphia: B. Miffline, 1850.

Reed, William Bradford. *Speech on the Presidential Question, Delivered Before the National Democratic Association, Philadelphia, September 4, 1860.* N.p., n.d.

Rowland, Mrs. Dunbar [Eron Rowland], ed. *Life, Letters, and Papers of William Dunbar of Elgin, Morayshire, Scotland, and Natchez, Mississippi: Pioneer Scientist of the Southern United States.* Jackson: Press of the Mississippi Historical Society, 1930.

Rudolph, Frederick, ed. *Essays on Education in the Early Republic.* Cambridge: Harvard University Press, 1965.

St. Mary's Hall, Twenty-Sixth Term. N.p., 1849.

Scharf, J. Thomas, and Thompson Westcott. *History of Philadelphia, 1600–1884.* 3 vols. Philadelphia: L. H. Everts, 1884.

Seward, Anna. *Letters of Anna Seward: Written between the Years 1784 and 1807.* 6 vols. Edinburgh: Constable, 1811.

Shryock, Richard H., ed. *The Letters of Richard D. Arnold, M.D., 1808–1876, Mayor of Savannah, Georgia, First Secretary of the American Medical Association.* Durham: Duke University Press, 1929.

"Southern Medical Students in Northern Medical Colleges—Once More." *Oglethorpe Medical and Surgical Journal* 3 (Jan. 1861): 268–69.

Staughton, William. *An Address, Delivered October, 1807, at Mrs. Rivardi's Seminary, on the Occasion of the Examination of the First and Middle Classes.* Philadelphia: Printed by Kimber, Conrad, 1807.

Stevens, Jonathan. "A Letter from New England, Part I." *Ladies' Repository* 12 (Jan. 1852): 21–24.

Taylor, Emily Heyward Drayton. *The Draytons of South Carolina and Philadelphia.* Philadelphia: Genealogical Society of Pennsylvania, 1921.

Tilghman, William. *An Eulogium, in Commemoration of Doctor Caspar Wistar, Late President of the American Philosophical Society Held at Philadelphia for Promoting Useful Knowledge.* Philadelphia: E. Earle, 1818.

———. "Report of the Historical and Literary Committee to the American Philosophical Society." *Transactions of the American Philosophical Society* new series 1 (1818): xi–xv.

Trapier, Paul. *Incidents in My Life: The Autobiography of the Rev. Paul Trapier, S.T.D. with Some of his Letters.* Edited by George W. Williams. Charleston: Dalcho Historical Society, 1954.

The Traveller's Tour through the United States: A Pleasing and Instructive Pastime, Perfomed with a Tetotum and Travellers. Philadelphia: Thomas T. Ash, 1835.

Trollope, Frances. *Domestic Manners of the Americans.* Edited by Donald Smalley. 1832; New York: Knopf, 1949.

Tyson, Job R. *Sketch of the Wistar Party of Philadelphia.* Philadelphia, 1846. Republished with additions by Henry C. Lea. Philadelphia, 1897 and 1976.

Wainwright, Nicholas B., ed. *A Philadelphia Perspective: The Diaries of Sidney George Fisher Covering the Years 1834 to 1871.* Philadelphia: Historical Society of Pennsylvania, 1967.

Wakefield, Priscilla. *An Introduction to Botany: In a Series of Familiar Letters, with Illustrative Engravings.* 6th ed. Philadelphia: Kimber & Conrad, 1811.

Waln, Robert Jr. [Peter Atall, pseud.]. *The Hermit in Philadelphia, Second Series. Containing some Account of Young Belles and Coquettes; Elegantes and Spoiled Children; Dandies and Ruffians; Old Maids and Old Bachelors; Dandy-Slang and Lady-Slang; Morning Visits and Evening Parties . . . .* Philadelphia: Moses Thomas, 1821.

Warren, Edward. *A Doctor's Experiences in Three Continents.* Baltimore: Cushings & Bailey, 1885.

Warren, John E. "Rambles in the Peninsula, No. IV." *Internatinal Magazine of Literature, Art, and Science* 1 (July 29, 1850): 136–38.

Webb, Samuel. *History of Pennsylvania Hall, Which Was Destroyed by a Mob, on the 17th of May, 1838.* Philadelphia: Merrihew & Gunn, 1838.

Webber, Mabel L., ed. "Josiah Smith's Diary, 1780–81." *South Carolina Historical and Genealogical Magazine* 33 (Jan. 1932): 1–28; (Apr. 1932): 79–116; (July 1932): 197–207; (Oct. 1932): 281–89; 34 (Jan. 1933): 31–39; (Apr. 1933): 67–84; (July 1933): 138–48; (Oct. 1933): 194–210.

Wharton, Anne Hollingsworth. *Genealogy of the Wharton Family of Philadelphia, 1664 to 1880.* Philadelphia: Printed by Collins, 1880.

Willson, Joseph. *Sketches of the Higher Classes of Coloured Society in Philadelphia.* Philadelphia: Merrihew & Thompson, 1841.

Wister, Fanny Kemble, ed. "Sarah Butler Wister's Civil War Diary." *Pennsylvania Magazine of History and Biography* 102 (1978): 271–327.

Wister, Owen. *The Philadelphia Club, Being a Brief History of the Club for the First Hundred Years of Its Existence, Together with a Roll of Its Officer and Members to 1934.* Philadelphia: Privately printed, 1934.

Wylie, Lollie Belle, ed. *Memoirs of Judge Richard H. Clark.* Atlanta: Franklin Printing, 1898.

Young Ladies Seminary. Philadelphia, [1843].

Secondary Sources

Allgor, Catherine. *Parlor Politics: In Which the Ladies of Washington Help Build a City and a Government*. Charlottesville: University Press of Virginia, 2000.

Amos, Harriet B. *Cotton City: Urban Development in Antebellum Mobile*. University: University of Alabama Press, 1985.

Anbinder, Tyler. *Nativism and Slavery: The Northern Know Nothings and the Politics of the 1850s*. New York: Oxford University Press, 1992.

Anderson, Benedict. *Imagined Communities: Reflections on the Origins and Spread of Nationalism*. Rev. ed. London: Verso, 1991.

Aron, Cindy S. *Working at Play: A History of Vacations in the United States*. New York: Oxford University Press, 1999.

Ayers, Edward L. *Vengeance and Justice: Crime and Punishment in the Nineteenth-Century American South*. New York: Oxford University Press, 1984.

Baatz, Simon. "Philadelphia Patronge: The Institutional Structure of Natural History in the New Republic, 1800–1833." *Journal of the Early Republic* 8 (1988): 111–38.

Baptist, Edward E. *Creating an Old South: Middle Florida's Plantation Frontier before the Civil War*. Chapel Hill: University of North Carolina Press, 2002.

Bacon, Margaret Hope. "Lucretia Mott: Pioneer for Peace." *Quaker History* 82 (1993): 63–79.

Baltzell, E. Digby. *Philadelphia Gentlemen: The Making of a National Upper Class*. 1958; New Brunswick, N.J.: Transaction Press, 1992.

Barlow, William, and David O. Powell. "A Dedicated Medical Student: Solomon Mordecai, 1819–1822." *Journal of the Early Republic* 7 (1987): 377–97.

Beckert, Sven. *The Monied Metropolis: New York City and the Consolidation of the American Bourgeoisie, 1850–1896*. Cambridge: Cambridge University Press, 1993.

Bell, Malcolm, Jr. *Major Butler's Legacy: Five Generations of a Slaveholding Family*. Athens: University of Georgia Press, 1987.

Bell, Whitfield J., Jr. "The Scientific Environment of Philadelphia, 1775–1790." *Proceedings of the American Philosophical Society* 92 (1948): 6–14.

Bernstein, Iver. *The New York City Draft Riots: Their Significance for American Society and Politics in the Age of the Civil War*. New York: Oxford University Press, 1990.

Blainey, Ann. *Fanny and Adelaide: The Lives of the Remarkable Kemble Sisters*. Chicago: I. R. Dee., 2001.

Bleser, Carol, ed. *In Joy and Sorrow: Women, Family, and Marriage in the Victorian South, 1830–1900*. New York: Oxford University Press, 1991.

Blight, David W. "'For Something beyond the Battlefield': Frederick Douglass and the Struggle for the Memory of the Civil War." *Journal of American History* 75 (1989): 1156–78.

Blumin, Stuart M. *The Emergence of the Middle Class: Social Experience in the American City, 1760–1900*. Cambridge: Cambridge University Press, 1989.

Bonner, Thomas Neville. *American Doctors and German Universities: A Chapter in International Intellectual Relations, 1870–1914.* Lincoln: University of Nebraska Press, 1963.

Bowie, Lucy Leigh. "Madame Grelaud's French School." *Maryland Historical Magazine* 39 (1944): 141–48.

Branson, Susan. *These Fiery Frenchified Dames: Women and Political Culture in Early National Philadelphia.* Philadelphia: University of Pennsylvania Press, 2001.

Breeden, James O. "Rehearsal for Secession? The Return Home of Southern Medical Students from Philadelphia in 1859." In *His Soul Goes Marching On: Responses to John Brown and the Harper's Ferry Raid.* Edited by Paul Finkelman, 174–210. Charlottesville: University Press of Virginia, 1995.

———. "States-Rights Medicine in the Old South." *Bulletin of the New York Academy of Medicine* 52 (1976): 348–72.

Brown, Dona. *Inventing New England: Regional Tourism in Nineteenth-Century America.* Washington, D.C.: Smithsonian Institution Press, 1995.

Brownell, Blaine E., and David R. Goldfield. *The City in Southern History: The Growth of Urban Civilization in the South.* Port Washington, N.Y.: Kennikat Press, 1977.

Bunkle, Philida. "Sentimental Womanhood and Domestic Education, 1830–1870." *History of Education Quarterly* 13 (1974): 13–31.

Burt, Nathaniel. *The Perennial Philadelphians: The Anatomy of an American Aristocracy.* Boston: Little, Brown, 1961.

Bushman, Richard L. "American High Style and Vernacular Cultures." In *Colonial British America: Essays in the New History of the Early Modern Era.* Edited by J. R. Pole and Jack P. Greene, 345–383. Baltimore: Johns Hopkins University Press, 1984.

———. *The Refinement of America: Persons, Houses, Cities.* New York: Vintage, 1992.

Carson, Carey, Ronald Hoffman, and Peter J. Albert, eds. *Of Consuming Interests: The Style of Life in the Eighteenth Century.* Charlottesville: University Press of Virginia for the United States Capitol Historical Society, 1994.

Chambers, Thomas A. *Taking the Waters: Creating an American Leisure Class at Nineteenth-Century Mineral Springs.* Washington, D.C.: Smithsonian Institution Press, 2002.

Cheves, Langdon. "Izard of South Carolina." *South Carolina Historical and Genealogical Magazine* 2 (1901): 205–40.

———. "Middleton of South Carolina." *South Carolina Historical and Genealogical Magazine* 1 (1900): 246–51.

Chew, William Lloyd, III. "Life in France Between 1780 and 1815 as Viewed by American Travelers." Ph.D. dissertation, Eberhard-Karls University (West Germany), 1986.

Chinard, Gilbert. "The American Philosophical Society and the World of Science (1768–1800)." *Proceedings of the American Philosophical Society* 87 (1943): 1–11.

Clinton, Catherine. "Equally their Due: The Education of the Planter Daughter in the Early Republic." *Journal of the Early Republic* 2 (1982): 39–60.

Cohen, Lester H. *The Revolutionary Histories: Contemporary Narratives of the American Revolution.* Ithaca: Cornell University Press, 1980.

Cohen, Patricia Cline. *The Murder of Helen Jewett: The Life and Death of a Prostitute in Nineteenth-Century New York.* New York: Knopf, 1988.

Cook, Charles M. *The American Codification Movement: A Study in Antebellum Legal Reform.* Westport, Conn.: Greenwood Press, 1981.

Corner, George W. *Two Centuries of Medicine: A History of the School of Medicine, University of Pennsylvania.* Philadelphia: Lippincott, 1965.

Cox, Robert S. "Vox Populi: Spiritualism and George Washington's Postmortem Career." *Early American Studies* 1 (2003): 230–72.

Cramer, Clayton E. *Concealed Weapon Laws of the Early Republic: Dueling, Southern Violence, and Moral Reform.* Westport, Conn.: Praeger, 1999.

Davis, Richard Beale. *The Abbé Corrêa in America, 1812–1820: The Contributions of the Diplomat and Natural Philosopher to the Foundations of Our National Life.* Providence, R.I.: Gávea-Brown, 1993.

DeRosier, Arthur, Jr. "William Dunbar: A Product of the Eighteenth-Century Scottish Renaissance." *Journal of Mississippi History* 28 (1966): 185–227.

Doerflinger, Thomas M. *A Vigorous Spirit of Enterprise: Merchants and Economic Development in Revolutionary Philadelphia.* Chapel Hill: University of North Carolina Press for the Institute of Early American History and Culture, 1986.

Dowling, William C. *Literary Federalism in the Age of Jefferson: Joseph Dennie and the Port Folio, 1801–1812.* Columbia: University of South Carolina Press, 1999.

Drinkwater, L. Ray. "Honor and Student Misconduct in Southern Antebellum Colleges." *Southern Humanities Review* 27 (1993): 323–44.

Duffy, John. "A Note on Ante-Bellum Southern Nationalism and Medical Practice." *Journal of Southern History* 34 (1968): 266–76.

———. "Sectional Conflict and Medical Education in Louisiana." *Journal of Southern History* 23 (1957): 289–306.

Dusinberre, William S. *Civil War Issues in Philadelphia, 1854–1865.* Philadelphia: University of Pennsylvania Press, 1965.

———. *Them Dark Days: Slavery in the American Rice Swamps.* New York: Oxford University Press, 1996.

Elkins, Stanley, and Eric McKitrick, *The Age of Federalism: The Early American Republic, 1788–1800.* New York: Oxford University Press, 1993.

Ezell, John S. "A Southern Education for Southrons." *Journal of Southern History* 17 (1951): 303–27.

Farnham, Christie Anne. *The Education of the Southern Belle: Higher Education and Student Socialization in the Antebellum South.* New York: New York University Press, 1994.

Faust, Drew Gilpin. *Mothers of Invention: Women of the Slaveholding South in the American Civil War.* Chapel Hill: University of North Carolina Press, 1996.

———. *A Sacred Circle: The Dilemma of the Intellectual in the Old South.* Baltimore: Johns Hopkins University Press, 1977.

Fischer, David Hackett. *The Revolution in American Conservatism: The Federalist Party in the Era of Jeffersonian Democracy.* New York: Harper & Row, 1965.

Foner, Philip S. *Business and Slavery: New York Merchants and the Irrepressible Conflict.* Chapel Hill: University of North Carolina Press, 1941.

Ford, Lacy P., Jr. *The Origins of Southern Radicalism: The South Carolina Upcountry, 1800–1860.* New York: Oxford University Press, 1988.

Franklin, John Hope. *A Southern Odyssey: Travelers in the Antebellum North.* Baton Rouge: Louisiana State University Press, 1976.

Gallman, J. Matthew. *Mastering Wartime: A Social History of Philadelphia During the Civil War.* Cambridge: Cambridge University Press, 1987.

Gienapp, William E. "The Crime against Sumner: The Caning of Charles Sumner and the Rise of the Republican Party." *Civil War History* 25 (1979): 218–45.

Geffin, Elizabeth M. *Philadelphia Unitarianism, 1796–1861.* Philadelphia: University of Pennsylvania Press, 1981.

———. "Nativism and the Creation of a Republican Majority in the North Before the Civil War." *Journal of American History* 72 (1985): 529–59.

Glover, Lorri. "An Education in Southern Masculinity: The Ball Family of South Carolina in the New Republic." *Journal of Southern History* 69 (2003): 39–70.

Gorn, Elliot J. "'Good-Bye Boys, I Die a True American': Homicide, Nativism, and Working-Class Culture in Antebellum New York City." *Journal of American History* 74 (1987): 388–410.

———. "'Gouge and Bite, Pull Hair and Scratch': The Social Significance of Fighting in the Southern Backcountry." *American Historical Review* 90 (1985): 18–43.

Grant, Susan-Mary. *North over South: Northern Nationalism and American Identity in the Antebellum Era.* Lawrence: University Press of Kansas, 2000.

Greenberg, Irwin F. "Charles Ingersoll: The Aristocratas Copperhead." *Pennsylviania Magazine of History and Biography* 93 (1969): 190–217.

Grimsted, David. *American Mobbing, 1828–1861: Toward Civil War.* New York: Oxford University Press, 1998.

Gross, Walter Elliott. "The American Philosophical Society and the Growth of Science in the United States, 1835–1850." Ph.D. dissertation, University of Pennsylvania, 1970.

Hackney, Sheldon. "Southern Violence." *American Historical Review* 74 (1969): 906–25.

Hale, Matthew Rainbow. "'Many Who Wandered in Darkness': The Contest over National Identity, 1795–1798." *Early American Studies* 1 (Spring 2003): 127–75.

Haller, John S., Jr. *Medical Protestants: The Eclectics and American Medicine, 1825–1939.* Carbondale: Southern Illinois University Press, 1994.

Halttunen, Karen. *Confidence Men and Painted Women: A Study of Middle-Class Culture in America, 1830–1870.* New Haven: Yale University Press, 1982.

Hanger, George DeLancey. "The Izards: Ralph, His Lovable Alice, and their Fourteen Children." *Transactions of the Huguenot Society of South Carolina* 89 (1984): 72–83.

Hansen, Karen V. *A Very Social Time: Crafting Community in Antebellum New England.* Berkeley: University of California Press, 1994.

Harris, J. William. *Plain Folk and Gentry in a Slave Society: White Liberty and Black Slavery in Augusta's Hinterlands.* Middletown, Conn.: Wesleyan University Press, 1985.

Hartridge, Walter Charlton. "St. Domingan Refugees in Maryland." *Maryland Historical Magazine* 38 (1943): 110–30.

Hatch, Nathan O. *The Democratization of American Christianity.* New Haven: Yale University Press, 1989.

Hebert, Catherine A. "The French Element in Pennsylvania in the 1790s: The Francophile Immigrants' Impact." *Pennsylvania Magazine of History and Biography* 108 (1984): 451–69.

Hemphill, C. Dallett. *Bowing to Necessities: A History of Manners in America, 1620–1860.* New York: Oxford University Press, 1999.

———. "Middle Class Rising in Revolutionary America: The Evidence from Manners." *Journal of Social History* 30 (1996): 317–44.

Hendrick, Robert. "Ever-Widening Circle or Mask of Oppression? Almira Phelps's Role in Nineteenth-Century American Female Education." *History of Education* 24 (1995): 293–304.

Hirsch, Adam J. *The Rise of the Penitentiary: Prisons and Punishments in Early America.* New Haven: Yale University Press, 1992.

Hobsbawn, Eric, and Terence Ranger. *The Invention of Tradition.* Cambridge: Cambridge University Press, 1983.

Horsman, Reginald. *Race and Manifest Destiny: The Origins of American Racial Anglo-Saxonism.* Cambridge: Harvard University Press, 1981.

Howe, Daniel Walker. *The Political Culture of the American Whigs.* Chicago: University of Chicago Press, 1979.

Jaher, Frederick Cople. *The Urban Establishment: Upper Strata in Boston, New York, Charleston, Chicago, and Los Angeles.* Urbana: University of Illinois Press, 1982.

James, D. Clayton. *Antebellum Natchez.* Baton Rouge: Louisiana State University Press, 1968.

James, Edward T., et al, eds. *Notable American Women, 1607–1950: A Biographical Dictionary.* 3 vols. Cambridge: Harvard University Press, 1971.

Johnson, Charles J., Jr. *Mary Telfair: The Life and Legacy of a Nineteenth-Century Woman.* Savannah: Frederic C. Beil, 2002.

Johnson, Mary. "Madame Rivardi's Seminary in the Gothic Mansion." *Pennsylvania Magazine of History and Biography* 104 (1980): 3–38.

Johnson, Michael P. *Toward a Patriarchal Republic: The Secession of Georgia.* Baton Rouge: Louisiana State University Press, 1977.

Johnston, Norman J. "The Caste and Class of the Urban Form of Historic Philadelphia." *Journal of the American Institute of Planners* 32 (1966): 334–50.

Jones, Howard Mumford. *America and French Culture, 1750–1848.* Chapel Hill: University of North Carolina Press, 1927.

Jones, Russell M. "American Doctors and the Parisian Medical World, 1830–1840." *Bulletin of the History of Medicine* 17 (1973): 40–65.

Jordan, Dr. Ewing. "University of Pennsylvania Men who Served in the Civil War, 1861–1865." In *Alumni Register: University of Pennsylvania,* 15–17 (1915–17).

Kaestle, Carl F. *Pillars of the Republic: Common Schools and American Society, 1780–1860.* New York: Hill & Wang, 1983.

Kashatus, William C. "'Punishment, Penitence, and Reform': Eastern State Penitentiary and the Controversy over Solitary Confinement." *Pennsylvania Heritage* 25 (1999): 30–39.

Kasson, John F. *Rudeness and Civility: Manners in Nineteenth-Century Urban America.* New York: Hill & Wang, 1990.

Katznelson, Ira, and Margaret Weir. *Schooling for All: Class, Race, and the Decline of the Democratic Ideal.* New York: Basic Books, 1985.

Kerber, Linda K. *Women of the Republic: Intellect and Ideology in Revolutionary America.* New York: Norton, 1986.

Kett, Joseph F. *The Formation of the American Medical Profession: The Role of Institutions, 1780–1860.* New Haven: Yale University Press, 1968.

Kierner, Cynthia A. *Beyond the Household: Women's Place in the Early South, 1700–1835.* Ithaca: Cornell University Press, 1998.

Kilbride, Daniel. "Philadelphia and the Southern Elite: Class, Kinship, and Culture in Antebellum America." Ph.D. dissertation, University of Florida, 1997.

Koschnik, Albrecht. "Political Conflict and Public Contest: Rituals of National Celebration in Philadelphia, 1788–1815." *Pennsylvania Magazine of History and Biography* 118 (1994): 209–48.

Lane, George Winston. "The Middletons of Eighteenth Century South Carolina: A Colonial Dynasty, 1678–1787." Ph.D. dissertation, Emory University, 1990.

Lazerow, Jama. "Rethinking Religion and the Working Class in Antebellum America." *Mid-America* 75 (1993): 85–104.

Lespinasse, Julie de. *Lettres de Mademoiselle de Lespinasse: Ecrites depuis l'année 1773, Jusqu'á l'année 1776; Suivies de Deux Chapitres dans le Genre du Voyage Sentimental de Sterne, par le Même Auteur. . . .* Paris: L. Collin, 1809.

Lewis, Charlene M. Boyer. *Ladies and Gentlemen on Display: Planter Society at the Virginia Springs, 1790–1860.* Charlottesville: University Press of Virginia, 2001.

Lewis, Jan. *The Pursuit of Happiness: Family and Values in Jefferson's Virginia.* Cambridge: Cambridge University Press, 1983.

Lewis, W. David. *From Newgate to Dannemora: The Rise of the Penitentiary in New York, 1796–1848.* Ithaca: Cornell University Press, 1965.

Ludmerer, Kenneth M. *Learning to Heal: The Development of American Medical Education.* New York: Basic Books, 1985.

Margo, Robert A. *Wages and Labor Markets in the United States, 1820–1860.* Chicago: University of Chicago Press, 2000.

Margo and Georgia C. Villaflor. "The Growth of Wages in Antebellum America: New Evidence." *Journal of Economic History* 47 (1987): 873–95.

May, Henry F. *The Enlightenment in America*. New York: Oxford University Press, 1976.

McCardell, John. *The Idea of a Southern Nation: Southern Nationalists and Southern Nationalism, 1830–1860*. New York: Norton, 1979.

McKelvy, Blake. *American Prisons: A Study in American Social History prior to 1915*. Chicago: University of Chicago Press, 1936.

McPherson, James M. "Antebellum Southern Exceptionalism: A New Look at an Old Question." *Civil War History* 29 (1983): 230–44.

Melish, Joanne Pope. *Disowning Slavery: Gradual Emancipation and "Race" in New England, 1780–1860*. Ithaca: Cornell University Press, 1998.

Miller, Jacquelyn C. "An 'Uncommon Tranquility of Mind': Emotional Self-Control and the Construction of a Middle-Class Identity in Eighteenth-Century Philadelphia." *Journal of Social History* 30 (1996): 129–48.

Morrison, Michael A. "American Reaction to European Revolutions, 1848–1852: Sectionalism, Memory, and the Revolutionary Heritage." *Civil War History* 49 (2003): 111–32.

———. *Slavery and the American West: The Eclipse of Manifest Destiny and the Coming of the Civil War*. Chapel Hill: University of North Carolina Press, 1997.

Murray, C. Craig. *Benjamin Vaughan (1757–1835): The Life of the Anglo-American Intellectual*. New York: Arno Press, 1982.

Murrin, John M. "A Roof without Walls: The Dilemma of American National Identity." In *Beyond Confederation: Origins of the Constitution and American National Identity*. Edited by Richard Beeman, Stephen Botein, and Edward C. Carter, 333–48. Chapel Hill: University of North Carolina Press for the Institute of Early American History and Culture, 1987.

Myers, C. Maxwell. "The Rise of the Republican Party in Pennsylvania, 1854–1860." Ph.D. dissertation, University of Pittsburgh, 1941.

Nash, Margaret A. "Rethinking Republican Motherhood: Benjamin Rush and the Young Ladies' Academy of Philadelphia." *Journal of the Early Republic* 17 (1997): 171–91.

Newman, Simon Peter. *Parades and the Politics of the Street: Festive Culture in the Early American Republic*. Philadelphia: University of Pennsylvania Press, 1997.

Norwood, William F. *Medical Education in the United States before the Civil War*. Philadelphia: University of Pennsylvania Press, 1944.

O'Brien, Michael. *Conjectures of Order: Intellectual Life and the American South, 1810–1860*. 2 vols. Chapel Hill: University of North Carolina Press, 2004.

———. *Rethinking the South: Essays in Intellectual History*. Baltimore: Johns Hopkins University Press, 1988.

O'Brien and David Moltke-Hansen. eds. *Intellectual Life in Antebellum Charleston*. Knoxville: University of Tennessee Press, 1986.

Oleson, Alexandra, and Sanborn C. Brown, eds. *The Pursuit of Knowledge in the Early American Republic: American Scientific and Learned Societies from Colonial Times to the Civil War*. Baltimore: Johns Hopkins University Press, 1976.

Olsen, Christopher. *Political Culture and Secession in Mississippi: Masculinity, Honor, and the Antiparty Tradition, 1830–1860.* New York: Oxford University Press, 2000.

Ownby, Ted. *Subduing Satan: Religion, Recreation, and Manhood in the Rural South, 1865–1920.* Chapel Hill: University of North Carolina Press, 1990.

Pace, Robert F. *Halls of Honor: College Men in the Old South.* Baton Rouge: Louisiana State University Press, 2004.

Patrick, Leslie. "Ann Hinson: A Little-Known Woman in the Country's Premier Prison, Eastern State Penitentiary, 1831." *Pennsylvania History* 67 (2000): 361–75.

Pease, William H., and Jane H. Pease. *James Louis Petigru: Southern Dissenter, Southern Unionist. Studies in the Legal History of the South.* Athens: University of Georgia Press, 1995.

Pessen, Edward. "How Different from Each Other Were the Antebellum North and South?" *American Historical Review* 85 (1980): 1119–49.

———. *Riches, Class, and Power before the Civil War.* Boston: Heath, 1973.

Rasmussen, Ethel E. "Democratic Environment–Aristocratic Aspiration." *Pennsylvania Magazine of History and Biography* 90 (1966): 155–82.

Redlich, Fritz. "The Philadelphia Waterworks in Relation to the Industrial Revolution of the United States." *Pennsylvania Magazine of History and Biography* 69 (1945): 236–54.

Richards, Leonard L. *"Gentlemen of Property and Standing": Anti-Abolition Mobs in Jacksonian America.* New York: Oxford University Press, 1970.

Rio, Angel del. *La Mision de Don Luis de Onis en los Estados Unidos (1809–1819).* Barcelona: Novagrafik, 1981.

Roark, James L. *Masters without Slaves: Southern Planters in the Civil War and Reconstruction.* New York: Norton, 1977.

Rorabaugh, W. J. *The Alcoholic Republic: An American Tradition.* New York: Oxford University Press, 1979.

Rosengarten, J. G. *French Colonists and Exiles in the United States.* Philadelphia: Lippincott, 1907.

Rosner, Lisa. "Thistle on the Delaware: Edinburgh Medical Education and Philadelphia Practice, 1800–1825." *Social History of Medicine* 5 (1992): 19–42.

Rothman, David J. *The Discovery of the Asylum: Social Order and Disorder in the Early Republic.* Boston: Little, Brown, 1971.

Rotundo, E. Anthony. *American Manhood: Transformations in Masculinity from the Revolution to the Modern Era.* New York: Basic Books, 1993.

Rugemer, Edward B. "The Southern Response to British Abolitionism: The Maturation of Proslavery Apologetics," *Journal of Southern History* 70 (2004): 221–48.

Ryan, Mary. *Cradle of the Middle Class: The Family in Oneida County, New York, 1790–1865.* Cambridge: Cambridge University Press, 1981.

Scarborough, William Kaufmann. *Masters of the Big House: Elite Slaveowners of the Mid-Nineteenth Century South.* Baton Rouge: Louisiana State University Press, 2003.

Schafly, Daniel L., Jr., "The First Russian Diplomat in America: Andrei Dashkov on the New Republic." *Historian* 60 (1997): 39–57.

Schankman, Arnold. "William B. Reed and the Civil War." *Pennsylvania History* 39 (1972): 455–68.

Schloesser, Pauline E. *The Fair Sex: White Women and Racial Patriarchy in the Early American Republic.* New York: New York University Press, 2002.

Scott, Anne Frior. "The Ever Widening Circle: The Diffusion of Feminist Values from the Troy Female Seminary 1822–1872." *History of Education Quarterly* 19 (1979): 3–25.

Scott, Donald M. *From Office to Profession: The New England Ministry, 1750–1850.* Philadelphia: University of Pennsylvania Press, 1978.

Sears, John. *Sacred Places: American Tourist Attractions in the Nineteenth Century.* New York: Oxford University Press, 1989.

Seelye, John. *Beautiful Machine: Rivers and the Republican Plan, 1755–1825.* New York: Oxford University Press, 1991.

Sellers, Charles S. *Mr. Peale's Museum: Charles Wilson Peale and the First Popular Museum of Natural Science and Art.* New York: Norton, 1980.

Shackelford, George Green. *Jefferson's Adoptive Son: The Life of William Short, 1759–1848.* Lexington: University Press of Kentucky, 1993.

Shade, William G. *Democratizing the Old Dominion: Virginia and the Second Party System, 1824–1861.* Charlottesville: University Press of Virginia, 1996.

Shaffer, Arthur H. *To Be an American: David Ramsay and the Making of the American Consciousness.* Columbia: University of South Carolina Press, 1991.

Sheriff, Carol. *The Artificial River: The Erie Canal and the Paradox of Progress.* New York: Hill & Wang, 1996.

Shields, David S. *Civil Tongues and Polite Letters in British America.* Chapel Hill: University of North Carolina Press for the Omohundro Institute of Early American History and Culture, 1997.

Sinha, Manisha. *The Counterrevolution of Slavery: Politics and Ideology in Antebellum South Carolina.* Chapel Hill: University of North Carolina Press, 2000.

Siry, Steven E. *DeWitt Clinton and the American Political Economy: Sectionalism, Politics, and Republican Ideology, 1787–1828.* New York: Peter Lang, 1990.

Sklar, Kathryn Kish. *Catharine Beecher: A Study in American Domesticity.* New Haven: Yale University Press, 1973.

Smith-Rosenberg, Carroll. "Domesticating Virtue: Coquettes and Revolutionaries in Young America." In *Literature & the Body: Essays on Populations & Persons.* Edited by Elaine Scarry, 160–84. Baltimore: Johns Hopkins University Press, 1988.

———. "The Female World of Love and Ritual: Relations between Women in Nineteenth-Century America." *Signs* 1 (1975): 1–29.

Snay, Mitchell. "American Thought and Southern Distinctiveness: The Southern Clergy and the Sanctification of Slavery." *Civil War History* 35 (1989): 311–28.

———. *Gospel of Disunion: Religion and Separatism in the Antebellum South.* New York: Cambridge University Press, 1993.

Stansell, Christine. *City of Women: Sex and Class in New York, 1789–1860.* Urbana: University of Illinois Press, 1987.

Stephens, Lester D. "The Literary and Philosophical Society of South Carolina: A Forum for Intellectual Progress in Antebellum Charleston," *South Carolina Historical Magazine* 104 (2003): 154–75.

———. *Science, Race, and Religion in the American South: John bachman and the Charleston Circle of Naturalists, 1815–1895.* Chapel Hill: University of North Carolina Press, 2000.

Stetson, Sarah P. "The Philadelphia Sojourn of Samuel Vaughn." *Pennsylvania Magazine of History and Biography* 73 (1949): 459–74.

Stevens, Kenneth R. "The Webster-Ingersoll Feud: Politics and Personality in the New Nation." *HIstorical New Hampshire* 37 (1982): 174–92.

Story, Ronald. "Class and Culture in Boston: the Athenaeum, 1807–1860." *American Quarterly* 27 (1975): 178–99.

Stowe, Steven M. "City, Country, and the Feminine Voice." In *Intellectual Life in Antebellum Charleston.* Edited by Michael O'Brien and Davied Moltke-Hansen, 295–324. Knoxville: University of Tennessee Press, 1986.

———. *Intimacy and Power in the Old South: Ritual in the Lives of the Planters.* Baltimore: Johns Hopkins University Press, 1987.

———. "The Not-So-Cloistered Academy: Elite Women's Education and Family Feeling in the Old South." In *The Web of Southern Social Relations: Women, Family, and Education.* Edited by Walter J. Fraser Jr., R. Frank Saunders Jr., and Jon L. Wakelyn, 94–120. Athens: University of Georgia Press, 1985.

Taylor, William R. *Cavalier and Yankee: The Old South and American national Character.* New York: Braziller, 1961.

Thibaut, Jacqueline. "'To Pave the Way to Penitence': Prisoners and Discipline at the Eastern State Penitentiary, 1829–1835." *Pennsylvania Magazine of History and Biography* 106 (1982): 187–222.

Thornton, J. Mills, III. *Politics and Power in a Slave Society: Alabama, 1800–1860.* Baton Rouge: Louisiana State University Press, 1978.

Travers, Len. *Celebrating the Fourth: Independence Day and the Rites of Nationalism in the Early Republic.* Amherst: University of Massachusetts Press, 1997.

Todd, Jan. *Physical Culture and the Body Beautiful: Purposive Exercise in the Lives of American Women, 1800–1870.* Macon, Ga.: Mercer University Press, 1998.

Wainwright, Nicholas B. "The Loyal Opposition in Civil War Philadelphia." *Pennsylvania Magazine of History and Biography* 88 (1964): 294–315.

———. "Sidney George Fisher: The Personality of a Diarist." *Proceedings of the American Antiquarian Society* 72 (1962): 15–30.

Waldstreicher, David. *In the Midst of Perpetual Fetes: The Making of American Nationalism, 1776–1820.* Chapel Hill: University of North Carolina Press for the Omohundro Institute of Early American History and Culture, 1997.

Wallace, David Duncan. *History of South Carolina.* 3 vols. New York: American Historical Society, 1934.

Walters, Ronald G. "The Erotic South: Civilization and Sexuality in American Abolitionism." *American Quarterly* 5 (1973): 177–201.

Warner, John Harley. *Against the Spirit of the System: The French Impulse in Nineteenth-Century American Medicine.* Princeton: Princeton University Press, 1998.

———. "The Idea of Southern Medical Distinctiveness: Medical Knowledge and Practice in the Old South. In *Science and Medicine in the Old South.* Edited by Ronald L. Numbers and Todd C. Savitt, 179–205. Baton Rouge: Louisiana State University Press, 1989.

———. "Orthodoxy and Otherness: Homeopathy and Regular Medicine in Nineteenth-Century America." In *Culture, Knowledge, and Healing: Historical Perspectives of Homeopathic Medicine in Europe and North America.* Edited by Robert Jütte, Guether B. Risse, and John Woodward, 5–29. Sheffield: European Association for the History of Health and Medicine, 1998.

———. "A Southern Medical Reform: The Meaning of the Antebellum Argument for Southern Medical Education." In *Science and Medicine in the Old South.* Edited by Ronald L. Numbers and Todd L. Savitt, 206–25. Baton Rouge: Louisiana State University Press, 1989.

———. *The Therapeutic Perspective: Medical Practice, Knowledge, and Identity in America, 1820–1885.* Cambridge: Harvard University Press, 1986.

Wayne, Michael. *The Reshaping of Plantation Society: The Natchez District, 1860–80.* Baton Rouge: Louisiana State University Press, 1990.

Warren, Leonard. *Joseph Leidy: The Last Man Who Knew Everything.* New Haven: Yale University Press, 1998.

Weigley, Russell F., ed. *Philadelphia: A 300-Year History.* New York: Norton, 1982.

Wells, Jonathan Daniel. *The Origins of the Southern Middle Class, 1800–1861.* Chapel Hill: University of North Carolina Press, 2004.

Wertenbaker, Thomas Jefferson. *Princeton, 1746–1896.* Princeton: Princeton University Press, 1946.

Williams, Jack Kenny. *Dueling in the Old South: Vignettes of Social History.* College Station: Texas A & M University Press, 1980.

Wood, Gordon S. *The Radicalism of the American Revolution.* New York: Knopf, 1992.

Wood, Kirsten E. "'One Woman So Dangerous to Public Morals': Gender and Power in the Eaton Affair." *Journal of the Early Republic* 17 (1997): 237–75.

Wyatt-Brown, Bertram. *The House of Percy: Honor, Melancholy, and Imagination in a Southern Family.* New York: Oxford University Press, 1994.

———. *Southern Honor: Ethics and Behavior in the Old South.* New York: Oxford University Press, 1982.

Young, Jeffrey Robert. "Ideology and Death on a Savannah River Rice Plantation: Paternalism amidst 'a Good Supply of Disease & Pain.'" *Journal of Southern History* 59 (1993): 673–706.

Zuckerman, Michael. "Holy Wars, Civil Wars: Religion and Economics in Nineteenth-Century America." *Prospects* 16 (1991): 205–40.

———. "Tocqueville, Turner, and Turds: Four Stories of Manners in Early America." *Journal of American History* 85 (1998): 13–42.

Index